Brain Policy

Brain Policy
How the New Neuroscience Will Change Our Lives and Our Politics

ROBERT BLANK

GEORGETOWN UNIVERSITY PRESS / WASHINGTON, D.C.

Georgetown University Press, Washington, D.C.
© 1999 by Georgetown University Press. All rights reserved.
Printed in the United States of America.

10 9 8 7 6 5 4 3 2   1999

Library of Congress Cataloging-in-Publication Data

Blank, Robert H.
    Brain policy : how the new neuroscience will change our lives and
our politics / Robert H. Blank.
        p.   cm.
    Includes bibliographical references and index.
    1. Neurosciences—Political aspects—United States.
    2. Neurosciences—Social aspects—United States.
    3. Neuropsychology—Social aspects—United States.   I. Title.
    RC343.B53—1999
    612.8′2—dc21   98-46088
    ISBN 0-87840-712-X (cloth).
    ISBN 0-87840-713-8 (pbk.)

Contents

Preface

My interest in the policy issues surrounding the brain was initiated while I was a member of the Office of Technology Assessment Advisory Panel on Neuroscience Research from 1987 to 1991. As the only social scientist on the panel, I was introduced a broad range of topics that I had never considered in my earlier work in biomedical policy which focused on human reproduction and genetics. I was struck by the often dramatic policy implications of an array of research initiatives and applications on the brain, particularly neural grafting, genetic interventions, and powerful new psychotropic drugs. I was also struck by the relative lack of emphasis on preventive-oriented research such as neurotoxins in the workplace and environment.

Out of this experience I decided to examine more carefully how work in neuroscience reflected more generic biopolicy issues. I concluded that while intervention in the brain presents similar concerns to areas more studied such as human genetic intervention and organ transplantation, it has largely been ignored as a policy area. Therefore, this book was written to fill a gap that I feel requires urgent attention. Within the context of rapid advances in our knowledge of brain structure and function and the end of the Decade of the Brain, hopefully this book offers valuable insights into this new area of biomedical policy. Since I am in effect an informed layperson in neuroscience, the book has been written to be read by anyone who has an interest in the broader social implications of research and applications in the brain. Although it contains technical information, the emphasis is on understanding the policy context and ramifications of these remarkable developments.

I would like to thank the reviewers who reviewed the manuscript and offered valuable suggestions. Although the manuscript was strengthened considerably by their input, all remaining errors or omissions are mine alone. I would also like to thank Jill Dolby who typed the manuscript and my wife Mallory for her continued support.

1

Introduction:
Intervention in the Brain

No two areas of medical research have wider implications for the study of the human condition than molecular biology and neuroscience. Although the political ramifications of human genetic research have been well documented and widely analyzed over the past decade, and the social, legal, and ethical dimensions funded as part of the human genome project, there has not been the same systematic scrutiny given neuroscience. In light of the rapid advances in our knowledge of the structure and functions of the central nervous system (CNS) and its linkages to genetics, it is critical to examine the impact of this new understanding and its accompanying vast array of applications on human behavior, social institutions, and our perceptions of humanhood.

Social scientists who in general have tended to ignore both genetics and neuroscience research must become aware of developments in the study of the brain and its implications for society in the twenty-first century. As noted by Crick, "there is no scientific study more vital to man than the study of his own brain" (1979: 32). Although neuroscience may not be the final frontier of humans, the benefits and risks of new areas of intervention in the brain, as in the genome, require heightened attention across the social sciences. In the words of Elliott White:

> A systematic synthesis of the subject matters of the neurosciences and the social sciences seems premature; but at the same time the central role of the brain in accounting for human thought and action, as increasingly suggested by neurobiology, dictates that this role be fully acknowledged within the social sciences. (1992: 19)

The central theme of this book is that the brain represents an important new biomedical area that must be studied much like genetics, reproductive techniques, and organ transplantation. As such, brain policy and politics is a critical area of study for social scientists and ethicists. In fact, given the brain's centrality to human existence, it is argued that brain policy be considered a prototype for analysis of the social impacts of biomedical intervention. This book attempts to explicate the political issues surrounding the remarkable advances in neuroscience and to argue for intensified analysis of the advantages and disadvantages they carry in their wake.

DECADE OF THE BRAIN

In 1989 Congress passed Public Law 101-58, declaring that the decade beginning January 1, 1990, be designated the "Decade of the Brain." According to the joint resolution, an estimated 50 million Americans are affected each year by disorders and disabilities that involve the brain. Brain disorders number over 650 and range from stroke and injury to Alzheimer's disease, schizophrenia, drug abuse, and alcoholism. As a category, brain disorders are a leading cause of death and the most common and severe cause of social, economic, and psychological disability in the United States. Brain-related disorders account for the majority of long-term care costs and for more hospitalizations and prolonged care than almost all other diseases combined. On any given day, 12 percent of adults in the United States suffer from a mental disorder and many others will be touched by it.

It is estimated that the economic impact of brain-related disorders and injuries in the United States is over $500 billion annually, a cost of almost $2,000 per person (U.S. Congress, 1995: 5). In 1991 more than one out of seven dollars spent on health care went to combat brain diseases. Psychiatric diseases alone accounted for more than $136 billion, neurological disorders $103 billion, alcohol abuse $90 billion, and drug abuse over $70 billion. Moreover, these economic costs do not include the intangibles, including disruption to families and accompanying social problems. Certainly, brain disorders represent a major challenge to science and society and make this focus on the brain warranted and timely.

The "Decade of the Brain" is also timely in the context of rapid research advances in the identification of complex anatomical connections; understanding the biochemical molecular and genetic mechanisms that control brain structure and functions; the ability to measure and visualize human brain functioning during mental activity; and the capacity to monitor neural activity simultaneously in complicated networks of neurons. In addition to treating neural diseases and disorders, these innovations promise increasingly precise and effective means of predicting, modifying, and controlling individual behavior. Although some molecular biologists might disagree, Selkoe concludes:

> Neuroscience is probably the great frontier now in biomedical science. It might have been cancer and cardiovascular disease a few years ago and before that the infectious diseases. Almost everything that you read about science now that is really the most exciting development relates to figuring out how the brain works. (1995: 84)

Although our knowledge of the brain is yet rudimentary, already in the 1990s great strides have been made and many new initiatives are bound to move to fruition and provide answers concerning brain activity and dysfunctions (see Cohen 1995 for a status report). The technology boundaries at this point, then, seem limitless.

According to the Office of Technology Assessment (OTA), "An atmosphere of enthusiasm surrounds neuroscience," the area of interdisciplinary research focused on how the nervous system works and how it is affected by disease (1992: 5). The Society for Neuroscience grew from 1,100 members in 1970 to over 22,000 by 1996. The 1980s saw almost a 70 percent increase in the number of papers published in neuroscience and behavioral research. By 1990, at least twenty government organizations were supporting research on neuroscience, with total federal expenditure exceeding $1 billion. As promising developments are announced, even more research attention is being focused on the brain.

Historically, experimental and clinical interventions in the brain, however, have elicited controversy from many directions. One need only look at issues surrounding past innovations of frontal lobotomies, electroshock therapy, and abuses of psychotropic drugs to see the sensitivity of intervention in the brain. Although new advances promise considerable benefits in treating a wide array of mental disabilities and behavioral problems, like genetic technologies the revolution in brain science challenges social values concerning personal autonomy and rights, and for some observers it raises the specter of mind control and an Orwellian-type society.

IMPLICATIONS OF THE AGING POPULATION

One factor that is likely to accelerate the incidence of neurodegenerating diseases and related neural problems is the aging of the U.S. population. The percentage of the population 65 years of age and over is expected to climb from 12.2 percent in 1996 to 20 percent in the next three decades. By 2030, one in five Americans will be over age 65, but more important, the cohort with the highest increases will be that category of those aged 80 and over. At present, the most rapidly growing segment of the population is the cohort aged 85 and over. In the United States between 1960 and 1990 there was an increase of 233 percent in those aged 85 and over, compared to an increase of 39 percent in the total population.

> Although this age group made up just 1 percent of the total population and 10 percent of the elderly population in 1990, demographers are watching it carefully. The higher rates of disability and poverty among these people are likely to have a growing impact on the nation's families and health and social service systems. (Randall 1993: 2332)

Conservative estimates suggest that population of those aged 80 and over could exceed 15 million by 2050, five times the present number. Some demographers, however, put the figure substantially higher at 20 to 40 million.

These figures have significant bearing on the distribution of health care resources. At present, 40 percent of U.S. health care resources go to the 12 percent of the population 65 years of age and over. Average per capita spending is approximately four times higher for the elderly than the nonelderly, and, importantly, the rate of increase in spending for the elderly is nearly three times that for the nonelderly. In 1987 per capita spending in the United States averaged $745 for persons under age 19; $1,535 for those aged 19 to 64; $5,360 for those aged 65 and over; and $9,178 for those aged 85 and over (Waldo, 1989: 116–18).

Ironically, because of medical improvements and technologies that prolong life, chronic disease that requires frequent medical care has become an increasing drain on scarce medical resources. Persons who in earlier times would have died of an illness are often kept alive to suffer long-term decline in quality of life. However, the demand for such intervention will continue to increase in an aging population. Furthermore, because of the concurrence of multiple and often chronic conditions, the cost of prolonging life at older ages is higher than at younger ones, increasingly so since the introduction of antibiotics in the 1940s reduced the incidence of death from illnesses such as pneumonia.

The implications of these demographic trends are especially significant in terms of the expected increase in dementias and other neurodegenerating conditions. The most common form of dementia is Alzheimer's disease, which accounts for approximately two-thirds of all cases, although over seventy other disorders can cause dementia. Alzheimer's disease causes progressive deterioration of memory, intellect, language, emotional control, and perception. The disease is insidious, gradually progressing from subtle symptoms to almost complete mental deterioration. At the final stage of Alzheimer's disease, the brain shows degeneration of the frontal and temporal lobes, particularly in the hippocampus, the area involved in short-term memory. Victims in the advanced stages may be completely dependent on others for years or even decades, causing serious disruptions and a huge financial burden to the family. The likelihood of developing dementia of some type is estimated at less than 5 percent of the population below the age of 75, about 20 percent of those aged 75 to 84, and more than 40 percent of those aged 85 and over (Institute of Medicine, 1992: 53).

At present, an estimated 4 million Americans suffer from Alzheimer's disease. Given the aging of the population, it is projected that the number affected will reach 14 million in the next century (National Institute on Aging, 1993: 1). Although the financial costs now estimated to approach $100 billion annually—including medical costs, nursing home and home care costs, and lost productivity—are great, the human toll of Alzheimer's and other dementias, including AIDS-related dementia, are also significant. Patients with these diseases confront the inevitability that their personal identity will disintegrate, while their families face the despair of watching helplessly as their loved ones' minds and bodies fade over periods of up to twenty years.

In order to avoid the tremendous economic, social, and personal costs that these trends imply, the research priority on the etiology of neurodegenerative disorders, suspected environmental and genetic causes, and potential treatments must be heightened significantly. Savings through such research are enormous. For instance, it has been estimated that a five-year delay in the onset of Alzheimer's disease could cut health care spending by as much as $50 billion per year in the United States. Similarly, a five-year delay in the onset of stroke could save $15 billion and a comparable delay for Parkinson's disease, $3 billion (U.S. Congress, 1995: 3).

BRAIN INJURY AND MENTAL DISORDERS

The aging population trend is not the only factor increasing the stakes of neuroscience research. Traumatic brain injury affects approximately 2 million people a year in the United States. About half of those injuries result in at least a short-term disability. Of those sustaining head injuries each year, about 500,000 require hospitalization and 75,000 to 100,000 die within hours of the injury. Moreover, approximately 90,000 people each year are injured severely enough to suffer irreversible and debilitating loss of function. Many will have difficulty returning to a productive life and require years of rehabilitation, medical follow-up, and integrated community services at a rate of $25 billion annually (Traumatic Brain Injury Act, 1993).

CNS injuries are the leading cause of death and disability for persons in the 15-to-24-year-old cohort. The leading cause of traumatic brain injuries are motor vehicle accidents (30 percent), followed by household accidents (26 percent) (Forkosch et al., 1996: 1). A contributing factor in many cases is alcohol consumption, and firearm injuries are a common cause of such trauma among young males. Research targeted to brain and spinal cord injury, therefore, must be accompanied by substantial social research on prevention strategies.

There is also some evidence that medical procedures themselves may contribute to neurological disorders. One study of patients from the twenty-four leading heart surgery centers, for instance, found that heart bypass surgery can lead to serious mental impairment from strokes and other brain damage at rates considerably higher than previously expected. In that study, 6.1 percent of survivors of this procedure suffered adverse neurological complications ranging from stroke to deterioration of memory, concentration, or other intellectual function. Approximately half of these patients suffered from severe neurological disorders, including fatal strokes (Roach et al., 1996: 1857). Given that over 300,000 heart bypass procedures are done each year in the United States, a figure that is expected to multiply rapidly with the aging population, this study suggests that as many as 20,000 people a year will experience major neurological damage as a result of undergoing heart bypass surgery.

Mental disorders represent an increasingly difficult policy issue in the United States and other developed nations. Although some categories of mental disorders appear to be increasing in prevalence (such as mood disorders among younger people), the diagnosis of such disorders continues to climb. Moreover, social changes in how the mentally ill are treated and the expansion of rights for persons with mental disabilities have resulted in difficult decisions as to how to balance individual and societal interests.

Although there are many ways to classify mental disorders, major categories include schizophrenia, mood disorders, and anxiety disorders.

Schizophrenia is "arguably the worst disease affecting mankind" because it "assails thought, perception, emotion, behavior, and movement, distorting an individual's personal experience of life and crippling his or her ability to participate in society" (Office of Technology Assessment, 1992: 47). Schizophrenia affects approximately 1 percent of the population and is five times more prevalent in the lowest socioeconomic groups. Its onset in the teens or early twenties can be insidious, but the sufferer faces excruciating symptoms, including delusions, hallucinations, and paranoia. Since the 1950s the drug chlorpromazine has been administered to reduce symptoms, but it does not affect the delusional thinking that shapes the patient's worldview. Because schizophrenia is a very complex disorder, it is unlikely that there is a single cause, although heredity is involved to some extent.

About one in twelve persons will be affected by a major mood disorder at some point in their lives. The most common form of mood disorder is depression, which affects about 5 percent of the population. Major depression may entail a single episode and is characterized by a complete loss of interest or pleasure in activities and can lead to suicide. Bipolar disorder (manic-depression) is characterized by major mood swings alternating between manic and depressive episodes. Manic-depression has clear genetic linkages and an average onset during the mid-twenties. Although lithium has been used successfully to control the symptoms of bipolar disease, it continues to be a highly debilitating disease for almost 1 percent of the population.

Anxiety disorders are increasing in prevalence in our high-stress age of anxiety. One common form is obsessive-compulsive disorder, which affects about 2.6 percent of the population. Although there is a limitless variety of forms and many sufferers may be otherwise healthy, obsessive-compulsive behavior is capable of destroying lives. Certain drugs and head injuries can bring on compulsive behavior, and data indicate that there is a genetic predisposition for this disorder. An additional 2 percent of the population suffer from panic disorder: sporadic, inexplicable spells of severe panic. Panic attacks lead to heavy use of medical services and can significantly alter the sufferer's health.

Cumulatively, these mental disorders affect millions of sufferers and their loved ones on a daily basis. Often they are accompanied by alcohol or drug abuse, which magnifies the effects. Although much of our knowledge concerning the causes of these diseases is yet speculative, it is study of the biochemistry, anatomy, and activity of the brain that offers the keys to understanding what causes these disorders and how to treat them.

> Convergent data using multiple neuroscience techniques indicate
> that the neural mechanisms of mental illness can be understood as
> dysfunctions in specific neural circuits and that their functions and
> dysfunctions can be influenced by a variety of cognitive and pharma-
> cological factors. (Andreasen, 1997: 1586)

The major challenge is to explicate why precisely these disorders of the
mind that arise in the brain occur and how to alter the molecular and
cellular factors that cause them.

Although we are only at the initial stages of understanding mental
disorders, it is clear that the biochemistry of the brain is critical to these
disorders and that activity in particular regions of the brain is altered by
this biochemistry. Schizophrenia, for instance, appears to block the brain's
receptor sites for dopamine in the prefrontal cortex region, which results
in abnormally low brain activity in that area. Panic disorders, likewise,
seem to be triggered by chemical imbalances that among other things
block action of the stress hormone epinephrine and produce asymmetry
between the right and left hemispheres in parts of the limbic system.
Obsessive-compulsive behavior and panic disorders are also linked with
abnormal action or inaction of various neurotransmitters (see Chapter 2)
including serotonin, dopamine, and norepinephrine.

Although presently hypotheses exceed evidence, it is likely that the
perplexing questions raised by mental illness are answerable. Fueled by
impressive advances in neurobiology, cognitive neuroscience, and genetics, we are likely not only to understand the biological causes of mental
disorders, but also to develop precise diagnostic tests and more effective
therapies for them (Gershon and Rieder, 1992: 127). These developments
will drastically improve the lives of millions of persons and lead to treatment of the disorders rather than suppression of their symptoms.

METHODS OF BRAIN INTERVENTION

The age of psychotechnology has arrived with the availability of a growing
arsenal of techniques for physical, chemical, and, potentially, genetic control of the brain. These innovations promise increasingly precise and
effective means of predicting, modifying, and controlling individual behavior. Although advances in brain science and technology are rapid,
much of the popular literature has oversimplified and exaggerated the
claims of presumed efficacy, thus heightening the fears of those persons
who see behavior control as a threat and creating unreasonable hope in

those persons who might benefit from it. Instead of eliciting a specific and predictable behavior, these interventions act upon the emotional state of the patient. Moreover, since no two human brains are alike, there is a large variation in effect from one person to the next, and even in the same person over time. This means that considerable uncertainty accompanies the application of any attempt to modify behavior. There is a significant hit-and-miss experimental element, despite our growing knowledge base. Also, since the brain is not neatly divided into discrete units conforming to categories of behavior, the effects of a particular intervention are not fully predictable.

The array of approaches for intervention in the brain include: (1) direct brain intervention techniques such as electrical stimulation and psychosurgery; (2) chemical, hormonal, or biological intervention; (3) genetic intervention; and (4) aversive and operant conditioning. The emphasis in this book is on the first three categories. For each approach, there are three application settings. The first is conducted on a voluntary patient who understands the potential risks and benefits of the procedure. Difficulties include determining when a patient is capable of informed consent and ensuring that procedural safeguards are provided to distinguish free choice from the second type of application—on a patient who consents under duress. Institutionalized patients are particularly vulnerable to strong pressure to "consent" to procedures, including electroshock and surgery (Sheldon, 1987: 549). The third and most problematic type of application is on a nonconsenting patient—either one who is incompetent to decide rationally or one who is forced against his or her will to undergo the "treatment." Often, the line between therapy and experimentation is blurred in all three applications, but the risk of solely experimental applications is highest in the third setting.

PUBLIC ATTITUDES TOWARD THE MENTALLY ILL

Mental disorders incur stigma from society. Surveys demonstrate that the mentally ill are regarded by many persons with fear, dislike, and distrust. Persons labeled as mental patients tend to be stigmatized and often avoided. Although mental illness has to some extent "come out of the closet," negative attitudes and ignorance of mental and psychiatric disorders are still commonplace. Although it is now less socially acceptable to admit fear or revulsion toward the mentally ill, recent survey data demonstrate that a sizeable majority of the population still feel that way. A 1991 survey conducted for the National Organization on Disability, for example, found that only 19 percent of the respondents felt very comfortable with a person who had a mental disorder (National Organization on Disability, 1991). This compares with 59 percent who were comfortable with a person in a wheelchair, 47 percent with a blind person, 39 percent with a deaf person, and 33 percent with a mentally retarded person. Highly publicized cases of violent mentally ill persons reinforces the poor perceptions of the public. For instance, while I was writing this section there was an announcement on the news that the public should be on the lookout for a psychiatric patient who escaped from the hospital and who was considered extremely dangerous.

Moreover, despite significant advances in medical knowledge about specific disorders and their treatment, ignorance remains high. Even some providers of mental health care are inadequately informed (see Ormel et al., 1990). This ignorance about mental disorders, as compared to more obvious physical ones, serves as a fertile ground for negative attitudes, though lack of knowledge alone cannot account for all the stigmatization that exists (OTA, 1992: 152). The attempted placement of treatment and housing facilities for the mentally ill in communities is often met with intense opposition. A 1990 survey found that this opposition increases from communities with higher income and education levels. Even mental health care professionals have been found to exhibit negative feelings toward individuals with mental disorders, particularly those with severe and persistent conditions (Dichter, 1992). These reactions to mental disorders are more similar to those toward drug addiction or ex-convict status than toward cancer or other medical disorders which themselves are stigmatized (Link et al., 1992). As noted earlier, studies suggest that a sizeable proportion of the public believes that mental illness is linked to violent behavior, a belief often reinforced by a minority of mentally ill individuals who display unusual or threatening behavior (Monahan, 1992).

The stigma attached to a mental disorder has significant social implications for public policy and for recognition and treatment of the problem. Affected individuals and their families suffer, and many family members are uncomfortable talking about the problem, thus leading to isolation and guilt (Lefley, 1992; Wahl and Harman, 1989). This response is reinforced because family members are often viewed as the agents of mental illness through psychoanalytic theories, for example, a schizophrenogenic mother who warps the psyche of her child. Individuals with mental disorders and their families may therefore avoid seeking treatment so as to prevent stigmatization. Moreover, believing that other people view them negatively, people with mental disorders react in a negative fashion and avoid treatment (Farina et al., 1992).

In addition to the impact of discrimination on persons identified as being mentally ill, the stigma attached to mental disorders has clear public policy implications. Public financing of treatment, housing, employment, and research is inadequate. Relative to their social costs, research for mental disorders is significantly underfunded as compared to AIDS, cancer, and heart disease (OTA, 1992: 153). According to the Interagency Task Force on Homelessness and Severe Mental Illness:

> Stigmatization, fear and mistrust regarding people with severe mental illnesses . . . are commonplace in our Nation. Such reactions influence both the direct responses of community members to these individuals as well as the development of local, state and Federal policies affecting them. (Interagency Council on the Homeless 1992)

The depth of the stigma, combined with the complexity of mental illness and the lack of clear and direct treatment regimes that are more common with physical disorders, makes any policy relating to mental disorders most controversial, especially when combined with genetic screening and testing.

POLICY ISSUES OF BRAIN INTERVENTION

Although some of the issues raised by neuroscience are unique, at their base they are similar across all segments of biomedicine. There are three policy dimensions that are central to all areas of biomedicine. First, decisions must be made concerning the research and development of the technologies. Because a substantial proportion of medical research is funded either directly or indirectly with public money, it is important that public input be included at this stage. The growing emphasis on forecasting and assessing the social as well as technical consequences of biomedical technologies early in research and development represents one means of incorporating broader public interests.

The second policy dimension relates to the individual use of technologies once they are available. Although direct government intrusion into individual decision making in health care has, until recently, been limited, the government does have at its disposal an array of more or less implicit devices to encourage or discourage individual use of technologies. These devices include tax incentives, provision of free services, and education programs. Although conventional regulatory mechanisms may generally be effective in protecting potential users or targets of new neuroscience applications, it is critical that their effectiveness be assessed and they be modified accordingly. Biomedical policy has a special importance in contemporary politics because it challenges keenly held societal values relating to privacy, discovery, justice, health, and rights. Biomedical techniques involve the human body and, with it, a deep-seated expectation of privacy.

The third dimension of biomedical policy centers on the aggregate societal consequences of widespread application of a technology. For instance, what impact would the diffusion of artificial heart transplants have on the provision of health care? How would widespread use of reproductive techniques alter our view of children, of families, and of women? Adequate policy making here requires a clear conception of national goals, extensive data to predict the consequences of each possible course of action, an accurate means of monitoring these consequences, and mechanisms to cope with the consequences if they are deemed undesirable. Moreover, the government has a responsibility in ensuring quality control standards and fair marketing of medical applications.

It is argued here that as one of the most dynamic and consequential areas of biomedical research, neuroscience must be analyzed along with other biomedical developments in this context. Research initiatives, individual use, and aggregate social consequences of unfolding knowledge about the brain and the accompanying applications require close scrutiny just as in other areas of biomedicine. The need for such analysis is heightened because of the centrality of the brain itself to human behavior and thoughts.

It is likely that the rapid diffusion of ever more sophisticated means of intervention in the brain will intensify. As one of the last frontiers of modern medicine, intervention in the brain promises benefits to many patients suffering from an array of neurological and mental disorders and injuries. Given the inevitability of expanded strategies for exploration and therapy of the brain, however, it is important that the policy issues surrounding their application be clarified and debated before such techniques fall into routine use. Whether the intervention is effectuated by chemical, surgical, genetic, or computerized means, the policy issues surrounding each application are similar, although they vary of course by degree of risk, probability of irreversibility, uncertainty of result, and experimental status.

Despite increased preciseness of diagnosis and treatment, there will always be uncertainty as to the impact of any intervention on a specific individual because each person's brain is in some ways unique. Moreover, the complexity of the human brain assures side effects that can never be fully predictable. This concern over risks and uncertainties is especially critical when the techniques themselves are experimental. As in other areas of medicine, there has been a tendency to move quickly from experimental to therapeutic status in developing these innovations. Although in some ways this is understandable, given the vulnerability of patients who normally undergo these techniques, their very vulnerability warrants a cautious approach. For instance, with regard to neural grafting, some observers have concluded that in the absence of extensive primate studies, "patients are being subjected to procedures for which the long-term prognosis is unknown" (Stein and Glasier, 1995: 37).

The blurring of the line between experimental and therapeutic status and the uncertainties surrounding application of any intervention on a particular patient raise important concerns over the physician-patient relationship. This situation is complicated when the physician is also the researcher who needs subjects for a yet unproven procedure. Again, while this issue accompanies any area of medical research, the unique characteristics of the brain make it especially problematic with any neural invasive procedures.

Especially problematic in brain intervention is the issue of informed consent, a concept that is at the base of medical research and therapy. Because consent for treatment of brain disorders must come from the damaged organ itself, it differs from intervention in other organs. If the damage is severe enough to warrant a high-risk, irreversible intervention, is the person undergoing treatment really capable of exercising free, informed consent? If not, who can make such a monumental decision for that person, and under what circumstances? This issue is presently at the heart of coercive use of psychotropic drugs for the mentally ill and will now be expanded to high-risk procedures for persons with neurodegenerative and genetic disorders. Although this problem most likely can be dealt with through existing mechanisms to protect human subjects or patients, it is often confused in new areas such as neural grafting and gene therapy, where the line between experimentation and therapy is often nonexistent and where the experimentation is done in a clinical setting.

Many of the innovations in direct brain intervention such as neural grafting and gene therapy are likely to be extremely costly on a per case basis. Moreover, many of the most expensive interventions are likely to be concentrated in populations of advanced age such as Alzheimer's patients. This raises important allocation questions regarding whether such procedures can be justified as a use of scarce societal resources. Interestingly, no cost figures or even estimates have been offered for these techniques and the assumption seems to be that cost should not matter. Although this issue is applicable to all medical innovations, it is especially important given the uncertainties surrounding brain intervention and the population to be served by it. It is likely that these applications, if widespread, will require further shifting of health care resources to the elderly and will represent further competition for the shrinking Medicare budget. These interventions also raise questions as to who gets priority for the use of these procedures, whether insurance companies must pay for what are at best experimental treatments, and the extent to which the government can or should regulate their use.

Assessing Brain Intervention Techniques

Ultimately, all technological advances must be judged on how well they improve the health and well-being of the population of both the present and future. The impact of neuroscience applications on future generations will be significant and thus must be considered when we decide whether and how to use them. On the one hand, the full benefit of today's research will accrue largely to our children and their children. New treatments for mental and behavioral disorders and for neurodegenerating diseases will follow our heightened appreciation of the interaction of genes, neurons, and the environment. On the other hand, however, the burdens, risks, and problems that accompany our expanded capacity to intervene in the brain will also largely be borne by generations that follow.

The reliance on the medical model of health in the United States, with its heavy emphasis on technological fixes, has worked at the expense of preventive approaches. Another product of this unrealistic faith in technology has been the failure of comprehensive assessments of the social and economic costs of new techniques prior to their diffusion. The promised or initial benefits of an innovation, as filtered by our predisposition to technological solutions, historically have blinded us to the downside. Thus, there has been a tendency to identify negative consequences only after the technique has become firmly entrenched, thus freezing it in place. In each area of neuroscience, therefore, from gene therapy and neural grafting to drug therapy and virtual reality, steps must be taken to analyze each innovation's impact prior to routine use. In other words, we need an anticipatory policy context instead of a reactive one in which we respond only after problems reach a crisis level. The lack of anticipatory policy in neuroscience might be especially costly in the future if we adopt without question the interventions discussed in this book.

Technology Assessment

Although medical technology assessment (MTA) has increased in magnitude over the last several decades, most efforts continue to be flawed. The failure of such assessment efforts has less to do with the capabilities of the assessing agents or the methods used than it does with the value context underlying medicine as practiced in the United States. The constraints on MTA are political and social and reflect the high personal stakes inherent in the life-and-death issues surrounding medical technology. As a result, MTA rarely rejects new technologies. Moreover, in those few instances where assessments recommend that certain procedures or drug therapies be limited, they are largely ignored, and development and application have proceeded for the most part unfettered.

As will be discussed in Chapter 6, psychotropic drugs, like all other drugs, are tested under strict guidelines and protocols set by the Food and Drug Administration (FDA), although of late these controls have been relaxed considerably in response to demand for experimental drugs by AIDS patients. There is no analogous regulatory process for medical technologies, however. This inconsistency was acknowledged by the OTA in 1982. It found that while drugs and medical devices were adequately tested, emerging medical and surgical procedures were not. Unlike other substantive technological areas, the reimbursement system, rather than a regulatory agency, may be the prime candidate for assessment because coverage and payment decisions by the government have become important factors in the diffusion of medical technologies.

Most other countries have incentive structures that constrain the diffusion of new technologies which lack proof that they will either lower the cost of treatment or provide a significant improvement in benefits. Japan regulates diffusion through a national fee structure, whereas other countries accomplish this control by a combination of mechanisms designed to limit high-cost innovations. By contrast, in the United States the insurance-dominated incentive structure and high economic stakes of the medical entrepreneurs generally encourage rapid diffusion of costly new interventions. Therefore, though procedures such as neural grafting are unlikely to be approved for payment in most public-oriented health care systems unless they can be shown to be cost-effective options, in the private-dominated U.S. system they will move quickly to therapy status even if they promise only a marginal benefit to a few patients. Although the resulting proliferation of medical technology means that insured Americans have access to the latest innovations, in many cases the innovations are of unproved benefit and in some cases they may actually be dangerous to the patient.

Despite the fact that the OTA's critical judgment of the assessment process for medical procedures is over a decade old, its observations remain relevant today. No class of medical technology is adequately evaluated on a continuing basis for its social and economic implications. In spite of efforts at MTA described above, there is no single organization whose mission it is to ensure that medical and surgical procedures are fully assessed before their widespread use.

Furthermore, the synthesis phase of MTA continues to be weak at best. Research evidence regarding the safety, efficiency, and effectiveness of emerging technologies is seldom analyzed systematically and objectively. As evidenced by the recent expansion of coverage for heart and liver transplantations and funding of AIDS research and treatment, reimbursement and regulatory decisions continue to be under the heavy influence of the political climate and clearly reflect a value system mired in the technological imperative. "Moreover, the redirection of such resources to alternative (nonmedical) pathways to better population health is not contemplated and, in the absence of coordination with public regulation, will not occur (Lomas and Contandriopoulos, 1994: 255)."

The general failure of MTA to challenge assumptions of the medical model and technological imperative leads some observers to believe it is bound to fail. According to Callahan, the technology assessment (TA) movement is but another example of our misdirected faith in technology to fix complex social problems. TA lacks any real value framework by which to make judgments on the moral or social worth of different technological goals. It can assess relative efficacy and economic consequences, but it cannot help determine whether it is justifiable to bear those consequences (Callahan, 1990: 92). MTA, according to Morgall, has not moved beyond the status quo, and it is built upon the assumptions of the medical model. As a result, "the possibility of totally rejecting the technology in question is not really an option in most methods, which do not challenge the technology but rather take it as given" (Morgall, 1993: 189).

According to Callahan, MTA will remain a "useless (and expensive) exercise unless there is a willingness to engage in prospective assessment before technologies are introduced, and to *force* a discontinuation of the use of those technologies that are ineffective or only marginally effective, or effective but too expensive to find social justification" (1990: 99–100).

The strong preference of the public and leaders for more and more advanced biomedical interventions, however, makes any attempts to restrict their development politically unattractive. The burden these technologies may place on future generations and the negative consequences that might accompany them are thus minimized or absent from most assessments.

The complexity of the interaction between medical technology and values requires considerably more attention to the long-term power of the technologies to alter values, often in unanticipated directions. Much of the current MTA tends to be linear in nature, with little appreciation of the interrelationships and dynamics of medical technology, politics, and values. As a result, efforts at assessing medical technologies tend to assume a static value framework and thus underestimate technology's impact on public expectations and usage and vice versa. In light of the tremendous potential of neuroscience to alter our conception of the human mind and behavior, such an assumption is especially shortsighted.

What is needed at this time is to reevaluate these values that now constrain MTA and to set social priorities. Again, this is not a new idea. A 1976 OTA report stated that "macroalternatives" to each technology being assessed should be defined. It is critical to consider alternative strategies to solve the same medical problem in very different ways and the effect that the technology in question will have on the development and implementation of those alternatives.

> For example, in assessing a therapeutic technology, one might consider proposals for prevention of the disease in question. It would be legitimate, in this context, to ask how reasonable, feasible, or desirable these alternatives are and whether heavy investment in or implementation of the therapeutic technology would encourage, discourage, or complement their development and implementation (OTA, 1976: 52).

As applied to neuroscience, this would require a comparative analysis of the relative benefits of placing priority on technological interventions or, alternatively, preventive efforts focusing on environmental and behavioral factors.

THE BRAIN PARADIGM

The huge costs of neurological disorders and injuries, both monetary and personal, themselves justify placing heightened research priority on neuroscience research. In addition, the knowledge gained from this research promises to provide answers to questions concerning the relationship between the brain, the mind, and the body, and each of their interrelationships with disease as well. Studies of brain development and plasticity can help us understand childhood and adolescence and suggest improved strategies for learning and teaching.

Conversely, this knowledge raises troubling issues and challenges traditional beliefs and values concerning free will, autonomy, and the meaning of life. It also exacerbates policy conflicts surrounding informed consent, privacy, discrimination, stigmatization, and the power to modify or control behavior. Without doubt, the neuroscience research and applications that follow will shake the foundations of social thought, giving credence to some theories of human existence and undermining others.

Despite lingering debate at the extremes by adherents of either genetic or environmental deterministic models of human behavior, in general most observers agree that some combination of nature and nurture are crucial. Often the disagreement centers on the proportional contribution of each factor, for example, what percent of variation can be explained by genetics and what percent by environment. In most cases, these positions fail to appreciate the dynamic, interactive nature of the genetics/environment relationship in their quest to explain the influence in either-or terms. This neglect becomes even more obvious when the brain is put into the equation.

The interactive paradigm holds that the genes and the environment are reciprocally related and therefore can influence one another over time. More important, both the genotype and the environment act to produce a specific phenotypic expression that defines the individual. Although this joint action serves to explain individual variation, it is not possible to legitimately generalize individual variation to population differences, despite sexual, racial, and ethnic patterns in genotype and social environment.

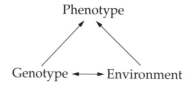

Though this first interactive model places nature and nurture in perspective, it fails to acknowledge what I argue here is the critical linkage factor, the brain. Even identical twins are unique individuals as a result of environmental influence mediated by the brain. Although maternal twins hold all genes in common and in some cases have frighteningly similar lives, the details of their neural connections are individually unique. In explaining behavior or capabilities, one cannot minimize the role of the brain or view it as an empty organism.

According to this expanded model, the brain is the key mediator of both genetics and the environment for the individual. As will be seen in later chapters, brain chemistry increasingly is being found as critical to our understanding of behavioral patterns, personality, and a range of individual capabilities. Neuroscience, therefore, offers an indispensable tool by which to explain why we are what we are and how we might make improvements on what we are. The brain provides a focus for analyzing the rich combination of genetic, environmental, and ultimately neural factors that define what we are. Unlocking the secrets of the brain is the key to explaining not only why humans differ from other species but also to how we vary among ourselves.

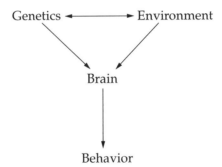

Conversely, any study of the brain's structure, function, and role without consideration of the genetic bases and environmental influences that shape its attributes would be remiss. The brain requires constant stimulation by the environment in order to develop. Without sensory input and intellectual challenges of a positive nature, full potential cannot be approached. Studies of infants increasingly demonstrate that the plasticity of the developing brain is remarkable but without either adequate genetic grounding or environmental stimulation its growth will be constrained. Thus while the brain is the great mediator, it too is dependent on the genotype of the individual and the environment within which the individual operates.

Ironically, just at the time when we are beginning to understand through neuroscience the full importance of the brain for human thought, capacity, and behavior, forces are at work to negate or at least weaken such claims. Criticisms of reductionism, of a chemical-based form of eugenics, are becoming increasingly vocal and vehement. Neuroscience, it is argued, threatens to unleash forces that shift social responsibility to individuals and that threaten institutions based on the environmental model. Moreover, others fear that neuroscience and cognitive research too easily dismiss the mind as nothing but chemical-electrical impulses and thus dehumanize us or somehow demystify humanhood. With these high stakes, neuroscience and the brain-centered paradigm will continue to elicit intense controversy in the coming decade.

THE FOLLOWING CHAPTERS

In order to understand the policy issues surrounding neuroscience, it is critical to have an adequate understanding of the structure and functions of the nervous system. Chapter 2 begins by providing an overview of this remarkable system, with an emphasis on basic anatomy and the mechanisms of the brain and related structures. The chapter then discusses the crucial fetal and early childhood stages of brain development and the importance of stimulation for neural growth. It also demonstrates the critical symbiotic relationship between brain, body, and environment, without which the brain cannot expand its potential. The chapter closes by looking at how the death of the brain represents the death of the person under contemporary public policy in the United States.

Chapter 3 presents several different theories of how the brain operates. The concepts of the brain as the "master juggler," as a computer, and as a modular organ are important to understanding the relationship of the brain to the mind. This relationship has fascinated philosophers for thousands of years and itself represents the power of the brain at its highest. What is consciousness, and can it be reduced to neuronal operations conducted by neurotransmitters? This chapter also briefly examines the current debate over cognitive research and the computer analogy of the brain. It then concludes that while the mind/brain (or mind/body) debate remains academically challenging, evidence from neuroscience clearly supports the view that the mind is nothing but what the brain does.

Chapter 4 explicates the connections between genetics and the brain and shows that though critical in establishing a template for brain activity, genetics fails to explain the infinite variation of neuronal connections. Current genetic research and applications targeted at understanding or manipulating brain function are also examined. The rapidly expanding capacity to intervene at the molecular level in the brain raises some of the most controversial issues concerning humanhood but also promises to alleviate a variety of neurological disorders.

Chapter 5 shifts emphasis to the question of what role the brain plays in explaining human behavior and the policy implications of such knowledge. It argues that one cannot ignore the brain's influence and that social scientists must include a neuroscience dimension in any theory of behavior. The chapter looks specifically at the role of the brain in violence and aggression, differences between the sexes, sexual orientation, and addictive behavior.

A widening range of techniques and strategies for intervention into the brain are quickly becoming available, often without adequate assessment. In addition to the genetic intervention techniques discussed in Chapter 4, many new physical, electronic, and chemical intervention techniques are emerging. Chapter 6 examines a variety of these techniques including virtual reality, their uses and potential abuses, and the policy issues they raise. In addition to issues of safety, efficacy, and consent, these interventions raise questions of resource allocation, the blurred lines between experimentation and therapy, and the dependence on methods that suppress symptoms rather than address causes.

Chapter 7 focuses on one procedure that has already engendered considerable public reaction. Neural grafting of fetal tissue into the brains of patients to treat neurodegenerating disorders offers promise of effective treatment but raises critical concerns over the growing demand for tissue from aborted fetuses. It is contended that all the public attention that was focused on the abortion issue diverted essential discussion of other important policy issues at the heart of neural grafting, issues common to the techniques discussed in Chapters 4–6. Chapter 7 concludes by demonstrating that although the debate over the use of fetal tissue has abated, many difficult issues surrounding neural grafting remain unresolved.

Neurotoxins are commonplace in the environment, but in most cases we lack clear evidence of their deleterious impact on the brain. There is a need for substantially more research and attention on neurotoxins in the workplace as well as in the community, but unfortunately our emphasis on the medical model precludes adequate funding for such research. In addition to these issues, Chapter 8 focuses attention on workplace hazards for pregnant women and on the continuing controversy about fetal health as it relates to neural development.

Finally, Chapter 9 returns to the broader questions of neuroscience and public policy. It argues for the early assessment of new intervention techniques and for the consideration of their potential economic costs and benefits. This chapter reiterates the need to construct workable preventive strategies based on our new knowledge of brain functions as alternatives to the conventional medical model. It also reiterates the argument for treating brain policy as a conceptual area on equal grounds with other areas of biomedicine and demonstrates the linkages between brain policy and other areas.

2

The Brain: Structure, Development, and Death

The brain has long been the subject of considerable speculation, myth, and misconception. Throughout history it has remained a mysterious and enigmatic entity: hidden within the skull, it represented a dark territory, little understood. Major technological developments in imaging the brain combined with leaps in knowledge about its functioning, however, have vastly expanded our understanding of its role in the last several decades. The evolving neuroscience perspective promises to help explain much about the biological bases of human behavior, consciousness, memory, language, and other attributes that make us what we are. Combined with research in molecular biology and other life sciences, our knowledge in neuroscience provides the key to understanding the foundations of human capacity as well as mental and behavioral dysfunction.

This chapter first summarizes current knowledge about the anatomy, development, and functioning of the central nervous system. It then describes the growing array of technologies that allow us to probe deeply into the recesses of the brain and examines the implications of those technologies. Finally, it examines the concept of brain death and its impact on the way we view life.

ANATOMY OF THE NERVOUS SYSTEM

One cannot appreciate the complexity of the subject matter of neuroscience without at least a rudimentary understanding of the structure of the human nervous system. Although all animals have nervous systems, the uniqueness of the anatomy of the human brain is crucial to explaining its functioning and the problems that arise when it fails to function normally. This first section provides a brief overview of the key components of the human nervous system, with special attention to those structural

23

characteristics which are essential to clarifying the centrality of the nervous system to our very definition of what it is to be a human.

The Nervous System

The nervous system is generally divided into two parts: the central nervous system (CNS) and the peripheral nervous system (PNS). The CNS includes the brain and spinal cord, and the PNS is composed of all the nerves, nerve cells, and specialized sensory receptors that lie outside the CNS. Sensory information about the world is brought into the CNS via the nerves and sensory receptors of the PNS. Moreover, the activation and regulation of activity in the glands, muscles, and organs are conveyed by the motor component of the PNS. In contrast, most information processing, decision making, and activity coordination is under the control of the CNS.

The Brain

The brain is a three-pound maze of nerve and tissue composed of the mushroom-shaped cerebral cortex and the brainstem, which is composed of hundreds of nuclei. A nucleus here is defined as a group of cells that share the same anatomical region. For every given function such as vision, hearing, and movement, a combination of nuclei and areas of the cerebral cortex act in a coordinated and highly synchronized manner.

The more scientists discover about the brain, the more complicated its functioning appears. Simple mechanistic conceptions of the brain have given way to the realization that it is the most complex structure in the known universe, constructed and maintained jointly by genes and experience (Fischbach, 1992: 48).

The anatomy of the brain and the intricacy of the connections among nuclei are exceedingly complex and thus able to combine, compare, and coordinate information from every area of the nervous system. "The brain determines the relevance of each bit of information, fits it in with all other information available, and decides what actions should or should not be taken to regulate bodily functions and to permit successful interaction with the external environment" (OTA, 1990: 30). The brain, then, is at the base of everything that makes up the existence of the organism, including consciousness itself.

The most striking part of the human brain is the cerebral cortex. The seemingly symmetrical hemispheres are covered by a laminated cor-

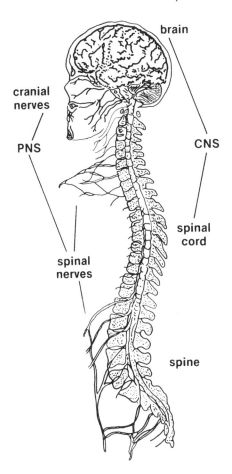

Figure 2.1   The Human Nervous System. (*Source:* C. Romero-Sierra, *Neuroanatomy, A Conceptual Approach* (New York, N.Y.: Churchill Livingstone, 1986.)

tex and are themselves subdivided by morphological and functional criteria. Roughly divided into four lobes on each side, the cerebral cortex contains numerous sensory-receiving areas, motor-control areas, language areas, and associated areas. The frontal lobes, for instance, are thought to be the seat of higher order thinking, which gives us the capacity to engage in abstract thinking and memory.

Although no category of cell or type of neural circuit is unique to the human brain, the major difference between humans and animals is the size of the cerebral cortex. The differential development of the human cerebral cortex is primarily accounted for by the vast increase in its surface area, which is folded to fit into a confined skull space. The efficiency of the highly convoluted surface of the cortex is remarkable. Although it is only about 2 millimeters thick, its surface area is about 1.5 meters. Accompanying the increase in the total number of neurons in the human brain is an expansion in the number and complexity of operations it can perform.

In comparison to other primates, the human brain has frontal lobes that are 200 percent larger. They encompass almost 40 percent of the total cortical area of the brain. The prefrontal areas are so uniquely human that animal experiments offer little insight into their activities. The last cortical areas to mature, the frontal lobes, are connected to almost every other part of the brain including the limbic system, which is responsible for expression of emotion. These extensive connections form the basis for the importance of the frontal lobe as an "integrator and regulator of brain function" (Restak, 1994a: 96).

Experience with patients who have suffered frontal lobe disease demonstrates how critical frontal lobes are to human thought and behavior. We are the only creatures capable of anticipating

> the future consequences of present actions; setting up plans and goals and working towards their achievement; balancing and controlling our emotions; and maintaining a sense of ourselves as active contributors toward our future well-being. These powers are diminished with frontal lobe disease. (Restak, 1994: 99)

The upper brain is surrounded by the cerebral cortex and contains structures such as the hypothalamus and the thalamus. Each of these structures is composed of a collection of nuclei with specific functions. The thalamic nuclei are primarily involved in analyzing and relaying sensory and motor information. Hypothalamic nuclei control behaviors such as eating, drinking, and sexual activity, and regulate the endocrine system. The basal ganglia nuclei help mediate movement, whereas the nuclei of the limbic system are involved in emotional behaviors. One limbic system nucleus, the hippocampus, is critical to learning and memory. Located at the base of the temporal lobe, the hippocampus is important for the consolidation of recently acquired information, or short-term memory.

Beneath the upper brain, the brainstem regulates essential bodily functions such as heart rate, blood pressure, and respiratory activity. Although the brainstem contains structures that regulate autonomic func-

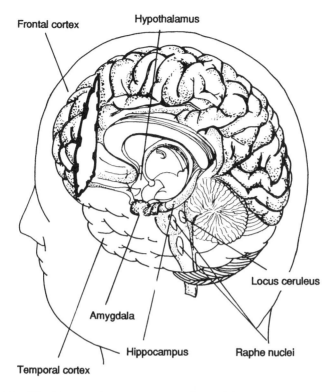

Frontal cortex

Hypothalamus

Locus ceruleus

Amygdala

Hippocampus

Raphe nuclei

Temporal cortex

Figure 2.2   The Human Brain (*Source:* Adapted from Lewis E. Calver, University of Texas, Southwestern Medical Center, Dallas, TX, 1992.)

tions that are not under a being's conscious control, this region is also an important junction for the control of deliberate movement. All the nerves that connect the spinal cord and the brain pass through the medulla at the lower end of the brainstem. In the pyramids of the medulla, many nerve tracks for motor signals pass from one side of the body to the other. As a result, the left hemisphere of the brain controls movement on the right side of the body and vice versa. The medulla is also the site of several pairs of nerves for organs of the chest and abdomen; for movements of the shoulder and head; and for swallowing, taste, hearing, and equilibrium.

The pons at the top of the brainstem serves as a bridge between the lower brainstem and the midbrain. Nerve impulses crossing the pons go to the cerebellum, where complex muscular movement is coordinated. Moreover, many nerves for the face and head that regulate movements of the eyeballs, facial expression, salivation, and taste have their origin in the pons. In combination with nerves of the medulla, nerves from the pons control breathing and one's sense of equilibrium.

The Spinal Cord

Extending downward from the brain stem is the spinal cord, which is composed of a central core of cells surrounded by pathways of axons. The cord is divided into thirty-one segments. Leaving and entering the spinal cord are nerves that bring sensory data into the CNS and transmit motor and activating information to the PNS. Dorsal nerve roots convey information into the spinal cord, while ventral nerve roots transmit information out to organs of the body and corresponding regions of muscles and skin in the limbs and along the trunk of the body. Also, some basic functions such as motor reflexes are organized and directed from within the spinal cord, although all these reflexes are modulated by the brain.

Peripheral and Autonomic Nervous Systems

The peripheral nervous system (PNS) is composed of all the sensory nerves coming into and the motor nerves leaving the spinal cord via the nerve roots, as well as some sensory and motor nerves that connect directly with the brain. The autonomic nervous system (ANS) is made up of the parts of the CNS and PNS that regulate visceral functions such as heart rate, digestion, and reflexive sexual activities. In the PNS the neurons of the ANS occur in specialized groups called ganglia.

The Endocrine System

The endocrine system is highly integrated with the CNS and is responsible for the production of hormones. Hormones are chemical messengers that are released into the bloodstream by endocrine glands. Among other purposes, they affect physical growth and development, sexual functioning, and emotional responses. The control center of the endocrine system is located in the center of the brain in the hypothalamus.

Beneath this region of the brain is the pituitary gland. The pituitary responds to signals from the hypothalamus by producing a variety of hormones that modulate the hormone secretion of other glands. The pituitary gland also produces several hormones with more general effects, including human growth hormone and dopamine, which inhibits the release of prolactin and acts as one of the many neurotransmitters. Imbalances in hormone secretions can cause both physical and mental disorders, including depression.

NEURAL COMMUNICATION

There are two major classes of cells in the nervous system. The first type, neurons or nerve cells, conduct all information processing through networks with other neurons. Each neuron consists of a cell body with long extensions called dendrites. Also projecting out of the cell body is a single fiber called the axon, which can extend for great distances to provide linkages with other neurons. All nerves in the body are bundled axons of many neurons conveying information to and from the CNS.

The second type of cells, glial cells, support neurons by carrying on critical chemical and physiological reactions and produce a variety of substances needed for normal neurological functioning. Some produce myelin, the fatty insulating material that forms a sheath around axons and speeds the conduction of electrical impulses. There is also evidence that oligodendrocytes inhibit the regrowth of damaged axons in the CNS, while Schwann cells support the regrowth of peripheral nerves. Other glial cells help regulate the biochemical environment of the nervous system and provide structural support for neurons. Astrocytes, for instance, function to remove degenerative debris in the nervous system and maintain the blood-brain barrier that protects the neurons from disruption by chemicals circulating in the blood.

Information in the nervous system is conveyed by chains of neurons, usually traveling in one direction, from the dendrites, through the cell body, and along the axon. This information is coded as an electrical-chemical message passed from one neuron to the next when its axon connects with a dendrite or cell body of another neuron at a synapse. The two neurons never actually make physical contact but remain separated by a small gap called the synaptic space. Furthermore, because the end of the axon and the dendrites of each neuron have many branches, the axon of any one neuron can form synapses with thousands of other neurons.

In the central nervous system, particularly in structures such as the cerebral cortex, the cerebellum, and the hippocampus, neurons possess long dendrites studded with thousands of synaptic spines. The cerebellar purkinje neurons possess on the order of 100,000 spines each, while the hippocampal neurons possess 5,000 to 10,000

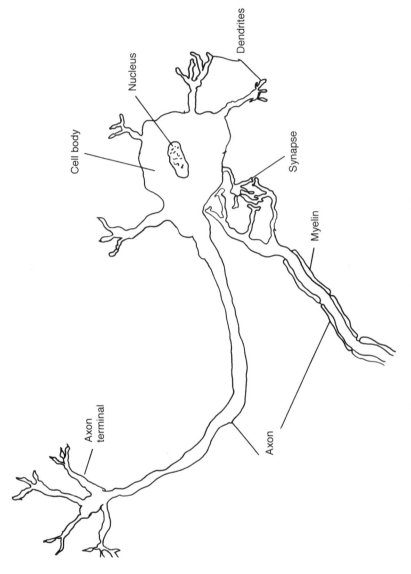

Figure 2.3 The Nerve Cell Structure (*Source*: OTA, 1992: 65.)

spines each. Every spine is usually contacted by one synapse. (Woj-towicz, 1996: 18)

Thus while the brain has approximately 200 billion neurons, the number of potential connections defies imagination at 1,000 trillion or more. Each axon can send and receive messages to and from thousands of other neurons.

Messages are sent when a neuron generates a nerve impulse, thus firing an electrical message along the axon toward the synapse. When the electrical message gets to the synapse, it causes the release of a chemical, one of scores of transmitters, from the axon into the synaptic space. These chemical neurotransmitters are released from tiny vesicles in the axon terminals and function like molecular keys that can unlock a channel in the membrane of the target dendrite and allow sodium ions to enter. A few of the critical neurotransmitters are dopamine, serotonin, epinephrine, norepinephrene, endorphin, acetylcholine, and gamma-aminobutyric acid (GABA). The neurotransmitter therefore acts as a messenger, crossing the

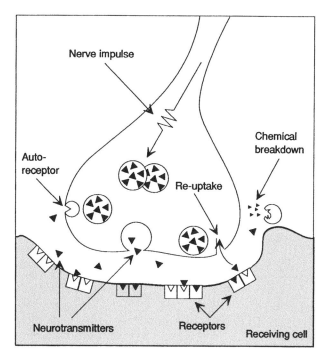

Figure 2.4   The Synapse (*Source:* OTA, 1992: 73.)

space and binding to the membrane of the next neuron. When this occurs, the charge between the inner and outer surface of the membrane is altered and the target neuron is briefly depolarized. The binding of the neurotransmitter to the neuron membrane either initiates, facilitates, or hinders an electrical message in that neuron, all happening in the space of a fraction of a second.

One yet unresolved question is how a neurotransmitter, a chemical compound unchanging in various parts of the body or brain, can be responsible for so many different effects. How can it produce such diverse results in its many different roles? The answer is likely to be found in the actions of the receptor sites, which are able to change their properties to a surprising degree and in effect determine how the neurotransmitter influences the cell at that specific moment. For instance, at least fifteen serotonin receptors have been identified; some are associated with depression, others with weight gain or hallucinations. As will be seen later, our understanding of brain function and dysfunction is likely to be dependent on our capacity to unravel the mechanisms and interactions of neurotransmitters at the separate receptor sites.

THE DEVELOPING BRAIN

During fetal development, the foundations of the mind are laid as billions of neurons form appropriate connections and patterns. Neural activity and stimulation are crucial in completing this process. (Shatz, 1992: 61)

The first signs of the nervous system in humans appear in the third week of embryonic development. The cells destined to become part of the CNS proliferate while migrating to their proper positions in the evolving brain. In order to arrive at the over 100 billion neurons that are the normal complement of a newborn baby, the brain must grow at an average rate of about 250,000 nerve cells per minute throughout the entire pregnancy (Institute of Medicine, 1992: 86). Following the period of proliferation and migration, the cells cease to multiply and begin to differentiate into the various types of neurons and glial cells that constitute the nervous system. By birth, the brain of the newborn has virtually all the neurons it will ever have, though the mass of the brain is about only one-fourth the size of the adult brain. The brain continues to grow because neurons grow in size and the numbers of axons and dendrites as well as the extent of their interconnections increase. The number of neuronal connections in newborns is estimated to be about 50 trillion, which expands to about 1,000 trillion in the first year after birth.

In fact, there are many more neurons in the developing nervous system than in the mature system. It has been estimated that about half of the cells initially generated die through competition for available synaptic sites and a matching of the number of neurons to the needs of the CNS. The survival-of-the-fittest process results in the survival of the neurons that have grown properly—approximately 200 billion neurons in the mature human brain. Although no new neurons are produced after the initial development phase, most glial cells maintain a capacity to replicate throughout adult life.

The process of maturation during the fetal period and childhood is not fully understood, but it has been found that in order to achieve the precision of the adult brain pattern, the brain must be stimulated (Shatz, 1992: 61). The number of neuronal connections can vary by 25 percent or more, depending on whether the child is exposed to a stimulus-rich or stimulus-impoverished environment. Development of the brain therefore is flexible and heavily dependent on the immediate environmental experiences of the child. Moreover, the process of axonal growth is thought to build from a genetically predetermined pattern of wiring that only grossly approximates the adult pattern. The growing axons are guided along these predetermined axonal pathways to their final destinations by neurotrophic and neurotropic chemicals. Neurotrophic, or nourishing, factors promote axonal growth and neuronal survival, whereas neurotropic factors provide a surface along which the axon can grow toward its target. Which target site or chemical address an axon searches for is thought to be genetically programmed within each neuron (OTA, 1990: 33).

Once the cells are in their proper positions, differentiation occurs, whereby the axons and dendrites begin to grow and forge thousands of synaptic contacts. Whereas the dendrites remain short and close to the cell body, axons become the long-distance transmitters, at times traversing distances that are the microscopic equivalent of miles and in other cases making all their synaptic contacts in close proximity to the cell. Although in humans most axons reach their target sites during gestation, the final configuration of synaptic contacts continues well after birth, when the brain experiences a second growth spurt as the axons and dendrites explode with new connections in response to the stimulation of the environment. By age 10, the brain's growth spurt slows, when synapse creation abruptly shifts to atrophy. Over the next few years, the weakest synapses are destroyed and the plasticity declines markedly.

The ultimate wiring of the brain with its trillions of synapses and connections is ironically the most complicated mechanism imaginable to the human brain. Its very complexity also makes it vulnerable to chemical assault, particularly in its sensitive developmental period (see Chapter 7).

IMAGING TECHNIQUES

Until recent decades, research on brain structure was based largely on post-mortem examinations of the brains of normal persons and those individuals who had suffered from mental disorders. New techniques that provide vivid images of living brains promise to greatly enhance our understanding of the relationship between the anatomy of the brain and psychological functioning. Increasingly sophisticated use of X-rays, radioactive tracers, and radio waves, combined with rapid advances in computerization, allow for noninvasive and safe investigation of the structure and functioning of the brain. The structure of the brain can be studied by computerized axial tomography (CAT), which uses computers to combine a series of X-rays to provide a precise picture of the brain.

Magnetic resonance imaging (MRI) can detect molecular changes in the brain when the individual is exposed to a strong magnetic field. MRI allows for clear and detailed images of brain activity and is used to detect structural abnormalities, changes in the volume of brain tissue, and the enlargement of cerebral ventricles in patients. The activity within particular regions of the brain can be analyzed to determine damage or malfunction and correlate it with behavioral manifestations. Echo-planar MRI (EPI) has significantly enhanced data obtained from standard MRI by using multiple, high-power, rapidly oscillating magnetic field gradients, higher speed hardware, and advanced image processing. Functional MRI (FMRI), which measures the increases in blood oxygenation that reflect a heightened blood flow to active brain areas, has higher resolution and faster speed than conventional techniques. Event-related FMRI promises to revolutionize brain research, according to Barinaga (1997).

Technologies specifically directed at the brain include measurement of electrical activity by enhancing conventional electroencephalographs (EEG) by computer analysis. Electrical activity can be measured while the patient performs particular cognitive or sensory tasks or is at rest, permitting investigators to observe changes in brain responses. Using knowledge of normal ranges, they are able to identify variations linked to particular mental disorders or behavioral problems. Magnetoencepha-lography (MEG) measures small magnetic field patterns emitted by the neuron's sonic currents and provides real-time resolution of the image to be studied.

Positron emission tomography (PET) and single-photon-emission computerized tomography (SPECT) are state-of-the-art imaging tech-niques that operate by creating computerized images of the distribution of radioactivity-labeled substances in the brain following their injection into the blood or through inhalation. As the radioactive substances move through the brain, investigators are able to visualize regional cerebral blood flow and glucose utilization as well as neurochemical activity. The more active a region is, the more blood will flow through it and the more glucose it will use. These techniques can measure abnormal activity in specific brain regions, in the whole brain, or in the normal asymmetry of activity between the two sides of the brain. Also, because PET scanning can use labeled drugs that attach to specific receptors, it is possible to identify the number and the distribution of receptor populations. For the first time in history, therefore, we have noninvasive techniques that allow for precise mapping of normal brain activity and for identifying variations from it that are related to specific behavioral manifestations.

These imaging systems are rapidly being followed by a new genera-tion of three-dimensional spatial imaging systems. This stereotactic im-aging combines a series of two-dimensional scans into a three dimensional virtual object. One such program at Brigham and Women's Hospital in Boston gives a doctor the ability to use a few keyboard commands to separate images of the various parts of the brain, making the cerebral cortex disappear to reveal the cerebrum in "fine detail with unprecedented clarity" (Hamit, 1994: 25). Similarly, the BrainSCAN Radiosurgery System provides three-dimensional imaging by correlating anatomic information from preexisting MRI with diagnostic data from CAT and angiography by means of automatic image fusion. Further advances in software are likely to match hardware improvements and provide even more remark-able and precise imaging of the brain.

Although still in the early stages, associations have been found between abnormal imaging patterns and specific mental disorders. For instance, some PET studies have found that decreased activities in the frontal cortex and limbic structures are associated with schizophrenia (Cleghorn et al., 1991). Moreover, EEG studies have observed a higher incidence of abnormal electrical activity in the brains of patients with schizophrenia (Levin et al., 1989). Likewise, studies using CAT and MRI have found that patients with bipolar disorders exhibit decreased cortical blood volume, indicating the possibility of structural abnormalities (Goodwin and Jamison, 1990). Data show that persons with bipolar disorder exhibit decreased cerebral blood flow and glucose utilization in the prefrontal cortex and a more general decrease in activity involving the whole cortex and the left frontal lobes. With these data derived from imaging technologies, our understanding of the structural and functional bases, as well as the myths, of mental disorders is being clarified.

THE BRAIN DEATH CONTROVERSY

Until recent decades, death was a relatively straightforward and unambiguous event. Death occurred at the moment of permanent cessation of respiration and circulation. Once the heart and lungs ceased functioning, they could not be restored. More important, once cardiorespiratory function ceased, brain function also ended. Advances in medical technology, which allowed for machine-regulated breathing and heartbeat even when the capacity to breathe spontaneously was irreversibly lost, however, combined with the demand for organs for transplant from brain-dead persons, made the conventional notion of death inappropriate by the late 1970s.

In 1981 the President's Commission concluded that "in light of ever increasing powers of biomedical science and practice, a statute is needed to provide a clear and socially-accepted basis for making definitions of death." Death would now be linked to cessation of brain function. Because the brain cannot regenerate neural cells, once the entire brain has been seriously damaged, spontaneous respiration can never return, even though breathing may be sustained by respirators or ventilators. The machines can maintain certain organic processes in the body, but they cannot restore consciousness or other higher brain functioning.

The present situation, in which all bodily functions need not cease when the heart stops pumping spontaneously, has led to the distinction between human life as a strictly biological existence and human life as an integrated set of social, intellectual, and communicative dimensions. Just what does it mean to be human? In recent decades, a reasonably strong consensus has developed that recognizes the possibility of social or cognitive death even though the human organism is kept alive biologically by artificial means. However, some observers remain opposed to the recognition of brain death, and others object to whole-brain death.

Still others who approve of the brain-death definition are uncomfortable with the dilemmas technology has created.

Whole-Brain Death

In one sense, the brain begins to die early in life as large numbers of redundant neurons are eliminated. Moreover, the normal aging process includes a gradual loss of sensory capacities. For instance, visual acuity declines linearly between ages 20 and 50 and exponentially after age 60. Depth perception declines at an accelerating rate after age 45. By age 80, speech comprehension may be reduced by more than 25 percent, due to extensive neuronal loss in the superior temporal gyrus of the auditory cortex (Ivy, 1996: 35).

The aging process, then, is naturally one of decline in the brain, with evidence suggesting a substantial decrease in neuron density as well as in absolute numbers of neurons in many of the brain regions, particularly the hippocampus, the subcortical brain regions, and the cerebellum (Selkoe, 1992: 135). As a result, "Total brain mass shrinks by approximately 5 to 10% per decade in the normal aged individual, leading to losses of 5% by age 70, 10% by age 80, and 20% by age 90. . . . The atrophy is most marked over the frontal lobes, although parietal and temporal lobes suffer considerable losses as well" (Ivy, 1996: 43).

In addition to the gradual process of brain-cell death that accompanies normal aging, an array of neurodegenerative diseases and neurotoxin exposure can cause further death among certain cell types and regions of the brain, as can injury, cancer, stroke, and other trauma (Morrison and Hof, 1997). High blood pressure, for example, has been found to shrink the size of the brain in elderly persons. There is also growing evidence that at least some of the changes that accompany aging are related to a decline in hormonal activity (Lamberts et al., 1997), with special emphasis recently placed on estrogen. Alcohol and drug abuse can intensify brain-cell death at all ages. The concept of whole-brain death, of course, refers not to this continuing process of cell death but rather to the complete cessation of brain activity and function, as measured by specific tests including EEG diagnostics.

The first major step toward defining brain death occurred in 1968, when rising concern by medical practitioners over how to treat respirator-supported patients led to the creation of the "Harvard criteria" for brain death (Ad Hoc Committee, 1968). These criteria, developed by a Harvard Medical School committee, focused on: (1) unreceptivity and unresponsiveness, (2) lack of spontaneous movements or breathing, and (3) lack of reflexes. Moreover, a flat EEG showing no discernible electrical activity in the cerebral cortex was recommended as a confirmatory test, when available. Before life-support systems could be terminated, all tests were to be repeated at least twenty-four hours later and must reveal no demonstrable change. These criteria, with some modifications and revisions due to new knowledge and diagnostic technologies, continue to serve as the standard medical criteria for determining brain death. Their publication led to the mobilization of considerable support for legislating policy standards of brain death and thus purportedly eliminated the uncertainties faced by hospitals and physicians.

Another force at work during this period was the emergence of organ transplantation techniques and the growing need for organs. For major organ transplants to be successful, a viable, intact organ was needed. The suitability of organs, especially the heart, lungs, and liver, for transplantation diminishes rapidly once the donor's respiration and circulation stop. Therefore, the most desirable donors are otherwise healthy persons who have died following traumatic head injuries and whose breathing and blood flow are artificially maintained until the removal of the organs. Although advocates of the brain-death criterion downplay the extent to which the demand for organs influenced this movement, transplant surgery did give the effort to define brain death a new urgency (Fox and Swazey, 1992). This reluctance to link the two developments is understandable, because the connection could imply that we accepted the revised definition of death only to facilitate use of the "dead" individual's organs. Certainly, organ-transplant facilities have sought and benefited from the legal clarity provided by statutes that define brain death. If the only rationale for this new definition of death was to facilitate successful organ transplants, however, support for brain-death determination would not have been as effective.

As with most matters of public health, the determination of death traditionally has been within the province of each state's common law. This dependence on the courts to determine and apply the criteria for death resulted in considerable uncertainty and a lack of consistency across jurisdictions. With patients often being transported across state lines for treatment, the lack of consistent policy produced confusion and potential for abuse. Two efforts to increase consistency resulted from this situation: many state legislatures enacted statutory standards, and a national standard has been largely implemented across the states.

In 1970 the Kansas state legislature became the first governing body to recognize brain-based criteria for determination of death. Within several years, four states passed laws patterned on the Kansas model. The Capron-Kass proposal (1972) offered the states a more succinct substitute, eliminating some of the Kansas model's ambiguity. At least seven states adopted the Capron-Kass model with minor modifications, while three others did so with more substantial changes. Two other model statutes (American Bar Association, 1975; National Conference of Commissioners on Uniform State Laws, 1978) were enacted by five and two states, respectively. The American Medical Association's 1979 proposal, which included extensive provisions to limit liability for persons taking actions under the proposal, was never adopted in any state. About ten states now have nonstandard statutes that often include parts of one or more of these models, whereas about twenty states do not have any clear statutory determinations of death.

This proliferation of similar yet variant models and statutes led the President's Commission to propose a "Uniform Definition of Death Act," which presently serves as the accepted standard definition. The act provides for:

> (Determination of Death.) An individual who has sustained either (1) irreversible cessation of circulatory and respiratory functions, or (2) irreversible cessation of all functions of the entire brain, including the brain stem, is dead. A determination of death must be made in accordance with accepted medical standards.

Before its presentation in the final report, this uniform law was approved by the American Bar Association, the American Medical Association, and the Uniform Law Commissioners as a substitute for their original proposals.

The commission recommended that uniform state statutes address general physiological standards rather than specific medical criteria or tests, since the latter continue to change with advances in biomedical knowledge and refined techniques. It concluded that "death is a unitary phenomenon which can be accurately demonstrated either on traditional grounds of irreversible cessation of heart and lung functions or on the basis of irreversible loss of all functions of the entire brain" (1981: 1).

Partial- or Higher-Brain Death

Although the whole-brain definition of death has become the accepted standard of practice in the United States, it has always been surrounded by controversy. Veatch (1993), for instance, argues that the whole-brain definition of death has become so qualified it can hardly refer to the death of the whole brain. For example, isolated brain cells continue to live and emit small electrical potentials measurable by EEG even though supercellular brain function is irreversibly destroyed. Whole-brain death also ignores spinal cord reflexes but requires cessation of lower brain stem activity, thus contradicting the definition that "all functions of the entire brain" be dead.

Other observers are even more critical of the current reliance on whole-brain death. Truog contends that the concept is fundamentally flawed, plagued by internal inconsistencies, and confused in theory and in practice (1997: 29). The specific tests for determination of death are inaccurate and do not always meet the required criteria. Furthermore, whole-brain death assumes that the brain is the integrating organ of the body whose functions cannot be replaced, even though intensive care units increasingly have become surrogate brain stems, replacing respiratory, hormonal, and other regulatory functions. Moreover, cardiac arrest is no longer inevitable upon cessation of entire brain function—it occurs only when it is allowed to occur. For Truog, "This gradual development of technical expertise has unwittingly undermined one of the central ethical justifications for the whole-brain criterion of death" (1997: 31).

One alternative to the whole-brain definition is partial- or higher-brain death. Higher-brain death is predicated on the assumption that particular areas of the brain are critical to the continuance of those functions which define us as humans. Personality, consciousness, memory, and reasoning require a functioning cortex. The higher-brain definition focuses on the loss of these characteristics. Thus cerebral death signals the death of the person when specific higher-brain functions cease, not all brain activity. This definition assumes that without consciousness, human life no longer exists.

Higher-brain death extends the definition to include patients in permanent vegetative states (PVS) and presumably anencephalic babies. Some observers would also include advanced stages of Alzheimer's disease under this definition, as well using newer imaging techniques such as FMRI and MEG to determine death. "As with irreversible coma, the parallel with brain death is plain. And a parallel policy of allowing such patients to perish would seem to be in order" (Churchland, 1995: 307). If it is these higher-brain functions that define us as humans, then partial-brain death, however narrowly defined, could be a more appropriate standard.

Reliance on higher-brain death itself has been criticized because of the difficulty of measuring with precision the loss of higher-brain functions and because it focuses on the end of personhood rather than on the death of the organism. It is argued that diagnostic tests for PVS at present are unreliable, as shown by anecdotal evidence of the recovery of some patients wrongly diagnosed as PVS. Some critics have also charged that if we accept the notion of higher- (or cerebral) brain death, we are vulnerable to sliding the slippery slope into finding as dead an ever widening range of marginally functional humans such as advanced dementia patients, as argued by Churchland (1995).

Truog suggests that higher-brain death is bound to remain the domain of philosophers rather than policy makers because of the implications of treating breathing patients as if they are dead (1997: 37). The public is unlikely to accept the burial or cremation of yet breathing humans or the use of lethal injections to terminate cardiorespiratory functions of the brain dead person prior to burial. This objection is congruent with the President's Commission rejection of partial-brain death on grounds that to declare dead a person who is spontaneously breathing yet has no higher-brain functions would too radically change our definition of death.

Despite these concerns, many observers feel that we are now at the stage technologically to move from whole- to partial- or higher-brain death. Veatch recommends changes in the wording to replace "all functions of the entire brain" with references to either higher-brain functions, cerebral functions, or his preferred "irreversible cessation of the capacity for consciousness" (1993: 23). Veatch would, however, incorporate a conscience clause that allows a person to choose through advance directive his/her preferred option based on personal, religious, or philosophical beliefs. Likewise, Emanuel (1995) proposes a "bounded-zone" definition of death whereby individuals are allowed to choose higher levels including PVS, but where traditional cardiorespiratory death would be the lower boundary for all persons.

Although higher-brain death has strong supporters, Truog rejects any movement in that direction and argues instead for a return to the cardiorespiratory standard. He contends that while whole-brain death served a useful transition function, brain death is no longer needed to justify withdrawal of life support, has little significance regarding resource allocation, and is not as crucial for organ transplantation as it was in the 1970s. He asserts that advance directives and sympathetic court decisions now allow withdrawal without reliance on brain death, although Truog fails to mention the vast inconsistencies of court decisions in this area.

Organ transplantation could be decoupled from brain death for Truog with a shift to the principles of consent of donor and no harm to the PVS patients and anencephalic infants. It is unlikely, I believe, that the courts or public would take this well because, as Truog notes, "The process of organ procurement would have to be legitimized as a form of justified killing, rather than just as a dissection of a corpse" (1997: 34). He is correct, however, that a return to cardiorespiratory death would eliminate the objections some groups have with brain death and would serve as the common denominator that it represented before technology altered the context of death.

Future Definitions of Death

Whatever standards for determining death are used, they will remain troublesome for some persons. Despite the widespread policy of brain death across the states, thousands of legally brain-dead persons are kept alive by artificial means, usually at the request or demand of the family. The difficulty of letting go, the false hope for a miracle, and the confusion of values resulting from the new technologies cause many persons to refuse to authorize unplugging the artificial life-support machines.

Under these circumstances, other difficult practical questions arise. Can third-party payers, including Medicare and Medicaid, refuse to pay for the care of a person who is legally dead? On what grounds can insurers justify such coverage? Veatch suggests the incorporation of a clause that standard health insurance providers not be required to cover medical costs to maintain any person who is "alive with a dead brain" (1993: 22). Can wills be probated in such a case? How can patients be protected from premature termination of helpful treatment under the guise of declaring death? What mechanisms are needed to maintain proper respect for the dignity of a brain-dead person when the various transplant teams need organs to save other patients who are brain-alive?

Although a few short years have powerfully transformed the meaning of death, many questions remain. Our very conception of what it means to be human is challenged by these rapid advances in medical technology. In light of the expanded knowledge of the brain and its functions, described more fully in the following chapters, it is likely that some variation of higher-brain death will become an accepted standard in the future. The growing financial and psychological burden on the living and the increase in neurodegenerating diseases due to an aging population will push us in that direction despite strong protests (Churchland, 1995: 307). More important, just as the active, living brain represents the center of personal identity and personhood, so the death of the brain that performs these identity-creating functions represents the death of the human person.

CONCLUSIONS

Simply stated, the brain is the central component to human life. In the absence of the functioning brain, human existence is impossible. The brain of each individual is an amazing, complex set of structures that we are only beginning to comprehend. Aided by recent developments in an array of highly sophisticated imaging techniques, however, neuroscience has rapidly built an impressive knowledge of brain structure and function. This knowledge, in turn, has enabled researchers to shed light on the biochemical bases of mental and behavioral disorders and the crucial role of neurotransmitters in explaining human behavior.

Policy issues arise throughout the life cycle, from early fetal development of the nervous system to the ultimate death of the brain. As we better understand brain function and dysfunction and the role of neurotransmitters in explaining behavior, the importance of the brain as a new biomedical area deserving of attention from policy analysts and social scientists will heighten. Ultimately, our very definitions of life and death are dependent on the findings of neuroscience. The following chapters shall draw out the emerging policy issues which demand attention.

3

The Brain, the Mind, and Consciousness

Despite the rapid expansion of the understanding of brain structure and function in recent years, we are still in our infancy in terms of developing a unifying theory of how the brain operates. As a result, considerable controversy surrounds neuroscience in at least four broad areas: the mind/brain distinction; the organization of the brain; the impact of genetics on the brain; and the role of the brain in determining human behavior. The latter two topics are discussed in Chapters 4 and 5, respectively. The issues concerning the organization of the brain and the mind/brain question are examined here.

Current arguments over how the brain operates and how it relates to the mind and to consciousness are not of recent origin. Instead, they extend back to the foundations of Western philosophy, when Aristotle rejected Plato's contention that a rational soul had its seat in the brain and instead advanced the cardiocentric view in which the brain is merely a cooling system for the body. Long after a quite universal rejection of Aristotle's heart-as-mind position, Descartes's dualistic theory of a separate brain and mind became the theory of choice of scholars and scholastics. Despite a growing agreement today that the mind is a construct which has outlived its usefulness and that the mind is what the brain does, the dualistic theory still has supporters who believe we cannot reduce thought and consciousness to biochemical interactions at the synapses.

Another assumption of many philosophers was that at birth the brain is a blank slate, or tabula rasa. This assumption gave society broad sway in teaching the citizen and shaping reality. Hobbes and Locke are among those philosophers who based their theories on this conception of the brain, a conception that is now challenged by evidence from neuroscience research.

44

A related long-standing controversy over the nature of the brain pits the holistic theories against those that emphasize localization and specialization of function. Under holistic doctrine, the brain functions as an indivisible whole; it is not reducible to its component parts. The holistic theory was dominant until the mid-nineteenth century. Although Gale's phrenology was soon discredited, his characterization of the brain as highly divisioned motivated empirical research that found localization of cerebral functions.

Paul Broca's celebrated finding that loss of speech (aphasia) was associated with lesions in the left frontal lobe revealed an asymmetry between the two hemispheres, which in turn contradicted holistic doctrine. Several years later, Carl Wernicke found that variations of aphasia were localized to lesions in the temporal lobe of the left hemisphere. Upon demonstrating that language was not a single function, but had at least three components carried out in different regions of the cortex, the rush for empirical research for localization was under way. The current representation of a modular brain actually combines some elements of earlier localization and holistic conceptions of the brain.

This chapter first discusses the modular brain theory and its implications for understanding the operations of the brain. It then analyzes the current attempts to characterize the brain as a computer and the issues they raise. Finally, attention is directed toward the continuing controversy over the relationship of the brain to what in the past was defined as the mind or the soul. It addresses the question as to whether thought and consciousness can ultimately be explained by neuroscience.

THE MODULAR BRAIN

According to holistic theories postulated by Descartes and many others, there is at some level a master site within the brain where all the separate components converge. This notion of a master control site, although intuitively attractive because it represents the I as a single entity, is not supported by our current knowledge of how the brain operates. The brain does not, and in fact cannot, act as a single integrated whole. Instead, very specific functions of the brain are highly localized, and these localized units (often termed modules) are linked together in a complex structure. For example, the neurons that allow our vision to differentiate straight lines are not the same as those that delineate curves.

The modular brain theory, however, transcends simple localization of function. What is most remarkable is that despite the division of labor, the brain has evolved structures that link these components together in predictable ways. Although specific functions are localized, all neurons and nuclei communicate with other modules. Multiple connections all operate simultaneously in parallel. This means that there is no cortical terminus, no master site or seat of consciousness. No one area holds sway over all others. All the separate modules do not report to a single executive center. For instance, while there is no one emotion center, the genesis and expression of emotions takes place in a constellation of groups of neurons or modules, or what Changeux terms "integration foci" (1997: 21).

The parallel processing capacity of the brain most likely has evolved as a survival mechanism, given the constraints of the brain structure. By computer standards, neurons act very slowly. Electrophysiological impulses in the brain travel at about 100 meters per second, about one-millionth the speed that electrical impulses travel over computer circuits. The brain is able to compensate for this relative slowness by using very many neurons simultaneously and in parallel and by arranging the system in a roughly hierarchical manner (Crick and Koch, 1992: 155). Any activity involves a wide network of numerous mutually interactive processes occurring simultaneously in many parts of the brain.

There are several areas where strong evidence supporting the modular theory has been uncovered. One of the most studied areas is language, where it has been found that the brain processes language by means of three interacting sets of structures. First, a large collection of nuclei in both the right and left cerebral hemispheres represents conceptual, symbolic interactions with the environment, mediated by sensory and motor systems. These functions categorize and organize objects, events, and relationships. Second, a smaller number of nuclei generally located in the left hemisphere represent individual sound units and syntactic rules for combining words. Finally, a third set of neural systems mediates between the first two. This third set can "take a concept and stimulate the production of word-forms, or it can receive words and cause the brain to evoke the corresponding concepts" (Damasio and Damasio, 1992: 89). Moreover, spoken and written comprehension occur in separate areas of the brain, with knowledge organized to include all modules operating simultaneously.

It makes sense that when groups of neurons must interact to carry out a specific function, they be localized in one region of the brain. Neurons in close proximity tend to receive similar input. Moreover, because of this proximity and the relatively short connections to nearby neurons, they allow for rapid interaction. From this standpoint, localization means that a division of labor is made most efficient by concentrating modules in one general region.

The fact that performance of simple functions is localized does not, however, negate the possibility that overall strategies for performing an integrative operation cannot be effectuated by combining different simple functions. Kosslyn and Koenig distinguish an integrative function (e.g., language) from the simpler component functions in arguing that we can have it both ways. Some functions are localized, but the brain also works as a whole to produce integrated functions that are not localized. The tasks of research, then, are to characterize what the functions are, which parts of the brain carry out each one, and how the functions work together (Kosslyn and Koenig, 1992: 12).

Early localization notions distinguished between the midbrain role in autonomic responses, the limbic system role in emotions and behavior, and the neocortex role in complex information processing, verbal language, and complex memory. Recent research shows this much too simple, with substantially more specialization throughout the neural system. Under the modular brain theory, the brain is a complex system of linked structures, with an emphasis on "linked." Localized assemblages of neurons have to communicate with each other or the entire system breaks down. In so doing, they actually change their structure and their interrelations (Masters, 1994: 7). The brain under this theory is an active, changing organ that requires constant intercommunication among distant neurons and nuclei in order to integrate the activity of these specialized modules.

This modular conception of the brain in conjunction with its adaptability in bypassing problem linkages and its parallel processing helps explain why some persons who suffer damage to one area of the brain may not exhibit the full extent of disability expected. Occasionally, a person with seemingly normal brain function is found upon autopsy to have only one hemisphere. It appears that when even such a major deficit occurs early in neural development, the brain's circuits are able to compensate for the absent hemisphere in spite of the localization. Usually, if damage comes later, for instance, upon removal of one hemisphere for treatment of intractable epilepsy (a hemispherectomy), the capacity for compensation is severely diminished.

For Wills, the key fact that distinguishes the human brain from that of animals is the "sheer amount of juggling our brains can do." The human brain, because of its redundancy in circuitry and parallel processing, is a "master juggler." "Many parts of the brain are involved in each function, and, although the various maps of the world that are found in different parts of the brain have different properties, one map may substitute for another to a surprising extent" (Wills, 1993: 276). If there was complete localization, of course, such substitutes could not occur. Damage to one area would destroy performance of that function. Although such destruction is possible in cases of major brain damage, in general the brain has remarkable flexibility. According to Restak (1994a), this demonstrates that the brain is opportunistic—it uses structure that may have evolved in one context to carry out different sorts of tasks.

Although specialization is a critical feature of the brain, there is a sharing of some functions by more than one system. To this extent, the brain is modular but only to a degree. Although the holistic theory is no longer appropriate, the brain is more complex than any localization theory alone could explain.

MEMORY AND REASON

Memory is central to human existence. At a most fundamental level, our personal sense of identity is found in our memory, our capacity to remember past experiences and build upon them. Memory is central to learning and to our ability to function at the most basic human levels. It is not surprising, therefore, that memory has been the focus of considerable neuroscience research, much funded to study the mechanisms of memory loss from dementias that deprive their victims of personal identity.

These studies of memory strongly reflect the modified modular theory of the brain. There is conclusive evidence that memory is not a single entity, but rather a process comprised of many essential components. Memory cannot be found in any single structure or location in the brain, though its components have been localized with increasing preciseness.

> Thus, there are no pictures stored in the brain, as was once thought. There are patterns of connections, as changeable as they are numerous, that, when triggered, can reassemble the molecular parts that make up a memory. Each brain cell has the capacity to store fragments of many memories, ready to be called up when a particular network of connections is activated. (Kotulak, 1996: 20)

There are many different dimensions and channels of memory storage that have been isolated. First, there are two separate channels of memory storage centered in different parts of the brain. Specific recall is centered in the temporal lobe and its connections to the limbic system. In contrast, habit formation, through which we remember how to perform skills, is a more diffuse system located primarily in the striatum. There is also evidence that the hippocampal system is involved in episodic memory that, over time (weeks or months), it transfers to the neocortex (Kandel and Hawkins, 1992).

In addition, there is a complementary relationship between two types of memory. Associative memory acquires facts and figures and holds them in long-term storage. However, such knowledge is of no value unless it can be brought to the forefront by working memory, itself a combination of different types of short-term memory. Working memory allows for short-term activation and storage of symbolic information and permits the manipulation of that information (Goldman-Rakic, 1992: 111). Working memory is the basic element in language, learning, thinking, and behavior. There is evidence that it is carried out in the prefrontal lobes, which also perform executive functions such as problem solving, planning, and organizing that require working memory.

There also have been identified separate neuronal circuits for spatial, object, and verbal working memories, although it remains debatable if there is one region that acts as a central processor for all working-memory information. Goldman-Rakic (cited in Wickegren, 1997a: 1581) believes there is not one such center but instead parallel systems, each with its own central processor. Recent studies using PET have found that working memories for facial features and their locations reside in separate regions of the prefrontal cortex and in separate sensory areas. It is expected that as research on memory expands, we will find even more divisions of labor by specific modules of neurons.

Repressed or False Memory?

This research might also lead to resolution of a current debate over memory with considerable legal ramifications, that over repressed memories. On the one hand, memory recovery of suppressed traumas such as child abuse has been used as critical evidence in prominent child abuse trials. Supporters argue that, unlike ordinary memory, traumatic-memory blocks are common when a person is somehow inhibited from completely processing and expressing the experience. Whitfield contends that most delayed or recovered memories are true and that there are no appropriate scientific studies or clinical trials that substantiate that these are imagined products of persons subjected to biased psychological therapy (1995: 66).

On the other hand, critics of recovered memory argue that the techniques used by some therapists to bring out blocked memories create false memories, the false-memory syndrome. Loftus and others have found that the brain is highly susceptible to such memory implants, which the person then believes are genuine. These researchers contend that memory fragments can be manipulated due to the malleability of memory and a human willingness to recall things that make sense. Young children and elderly people with short attention spans appear most vulnerable to false-memory syndrome. Researchers such as Loftus often appear as expert defense witnesses of accused child abusers. In contrast, their detractors argue that there is no evidence for false-memory syndrome. The answer to this acrimonious debate is likely to be found only in a deeper understanding of how human memory works and what neural networks are responsible for each type of memory.

Reason

Closely related to memory is the function of reason or thought. Kosslyn and Koenig (1992) see reason as the best example of an integrative function. Like memory, reason requires the orchestration of many component processes. In addition to memory subsystems, reasoning incorporates a host of processing subsystems, including those of perceptual encoding, imagery, action, and perceptual input. Also, like memory, the reasoning process assumes the presence of a decision system that coordinates all the others so that a specific goal can be met.

THE COMPUTER ANALOGY

One controversy that has sparked considerable debate among neuroscientists centers on the appropriateness of the brain-as-a-computer analogy. Cognitive science, a cross between psychology and computer science, has found it useful to reduce the brain function to that of a machine. This man-machine interface concentrates on formal operations and programs or subprograms of the brain instead of on nuclei and neurotransmitters. Even outside of cognitive science, however, it is now popular to use computer terminology such as hard-wiring, software, inputs, storage retrieval, and parallel processing to describe brain structure and functions.

This is understandable, given the ubiquitous nature of computer language in society, but it does raise questions as to how accurate and helpful the computer model is in describing the brain.

Cognitive science has developed rapidly and combines the perspectives of cognitive psychologists, specialists in artificial intelligence, computer scientists, and engineers. Adherents of the computer analogy believe it offers many advantages. First, it provides new ways to mimic the activity of complex networks of neurons, thereby allowing researchers to formulate more precise theories of brain function (see Hinton, 1992). Second, computer-aided scanning techniques allow these theories to be tested by monitoring the activity of a functioning brain and observing which regions are involved in specific cognitive activities. In combination, computer modeling and scanning provide the theory and the data needed to develop more explanatory models of brain activity.

According to Kosslyn and Koenig, cognitive neuroscience can improve our understanding of mental capacities, including memory, perception, and language, by "delineating component processes and specifying the way they work together" (1992: 3). Under this approach, the mind is nothing but what the brain does and what the brain does can be reduced to formal processes based on the computer models.

Although Hinton admits that we have a long way to go in clarifying which representations and learning procedures are used by the brain, "sooner or later computational studies of learning in artificial neural networks will converge on the methods discovered by evolution" (1992: 151). When that happens, he concludes, a lot of diverse empirical data about the brain will make sense.

Recent research, for instance, has explored the implications of one principle of neural computation (optimization) for the theory of grammar. For Prince and Smolensky (1997), optimization principles allow a variety of new insights into the structure of the language faculty when the relation between optimality in grammar and optimization in neural networks is considered. Another study combines statistical and probabilistic aspects of language, connectionist models, and the learning capacities of infants to study language acquisition and use. This study concludes that "research in developmental neurobiology and in cognitive neuroscience has begun to yield more direct and specific evidence as to how brains are structures and develop" (Seidenberg, 1997: 1602).

Although cognitive neuroscience has become a highly visible approach to the study of the brain, and without a doubt the computer is an invaluable tool to this end, many observers are highly critical of the presumption that the brain can be represented accurately by a computer analogy.

> Because it is so immensely flexible in the way it can juggle inputs, the brain does not resemble an ordinary computer in the least. The programming of the brain is so flexible that there are always many alternate pathways for data to be processed—or nearly always. (Wills, 1993: 270)

Changeux (1997) claims that the computer analogy is deceptive because, unlike the computer, the brain is capable of developing strategies of its own. Also unlike the computer, the brain makes no distinction between hardware and software. The brain is capable of building representations of its own in response to the outside world and of using them in its own computations, something that even the most powerful computers are unlikely to achieve.

Hooper and Teresi argue that the computer analogy fails because "[t]he brain is not really like anything except a brain" (1992: 13). They firmly reject the "reductionist dream" to simplify mental states to microcomponents of the brain. It is similarly wrong, they argue, to transform the brain into a "symbol cruncher," because all knowledge cannot be formalized as the cognitive scientists assume.

Critics seem to be especially concerned with what they view as the desire of more "extreme" researchers in artificial intelligence and perception to substitute computer chips for neurons. According to Hinton, these networks of artificial neurons are essential for mimicking the brain's learning process. "We construct these neural networks by first trying to deduce the essential features of neurons and their interconnections. We then typically program a computer to simulate these features" (1992: 145). Computer chips have already been developed that are capable of sensory processing. Others with memory-storing capacity are being developed. Synapses have been built into computer chips that perform "learning" at the rate of millions of times per second. Moreover, a new generation of chips promises to represent neurons with many thousands of synapses (Institute of Medicine, 1992: 140). Some cognitive scientists, buoyed by success in chess-playing computers, foresee the eventual development of a computer with consciousness.

Restak, however, terms as "nonsense" the possibility of anything approaching consciousness resulting from such an endeavor. He argues that even though computers can easily outperform the brain in handling mega amounts of data, they will never be able to tell us how the brain operates. Restak opposes pursuing the computer analogy on several grounds. First, he contends that one cannot use a vastly simpler level of organization like the computer to explain in entirety a more complex one (although it must be noted here that few cognitive scientists have suggested this was possible). Second, Restak points out that for all the posturing by cognitive scientists, all the insights about parallel processing and modular organization have come from basic research on the brain, not from computer modeling. The proper way to learn about the brain is to study the brain, because "the brain is organized according to the rules of neurophysiology, not computer science, and there is simply no way of getting around that fact" (Restak, 1994a: 177).

Moreover, the brain structure has evolved over millions of years, and therefore the brain is a unique product of a long and complicated process we cannot duplicate. The formative influences of development of the brain are difficult to identify, much less incorporate into a computer program. No matter how enticing the computer analogy may appear, brains are not biological computers. Our goal-seeking behavior and capacity to change goals midstream distinguishes us from computers, despite their tremendous advantages in processing. If the brain is not like a computer except in some obvious but limited ways, we are therefore unlikely to learn much on how the brain works by placing our confidence in the computer model. This should not preclude the use of computers whenever appropriate to aide in our quest to understand the brain, but it does mean that we might be better off by placing our highest priority on less dramatic but crucial basic neurological research.

THE END OF DUALISM?

> The overwhelming question in neurobiology today is the relation between the mind and the brain. (Crick and Koch, 1992: 15)

Although it is not within the realm of this book to explicate the debate over how the mind relates to the brain, it is important to examine how emerging knowledge in neuroscience will affect the debate. Is the mind a collection of mental processes, or is it a substance or spirit beyond the physical brain? Are there mental states independent of the physical brain, or is the existence of a conscious mind apart from the functioning brain only a product of fancy or of the ignorance of the full capacities of the brain? Moreover, if the mind and body are distinct, how do they interact?

These questions are as old as philosophy itself, and they are unlikely ever to be resolved.

Until recent times, the brain itself was largely ignored. Attention was directed to the ethereal, mysterious mind and to the waking mind or consciousness. Descartes's notion that the mind was an immaterial, extracorporeal entity expressed by the pineal gland was matched by Willis, who opted for the striatum as the seat of the mind. Dualism, the theory that the mind is separate from the brain or the body, continues to elicit support from some scientists and philosophers. Sir John Eccles, for example, believes that an immaterial soul or "ghost" is the essence of our conscious self. The mind is not reducible to the brain machinery, and the mind controls the brain through free will emanating somewhere in the supplementary motor area. Other recent dualist thinkers include David Chalmers (1996), who calls for "psycho-physical bridging laws" that cross the mind-body gaps. Roger Sperry suggests that while the mind's source is the brain, the mind cannot be reduced to neuronal activity, nor fully explained by neuroscience. The whole (the mind) is considerably greater than the sum of its parts (the physical brain structures).

Although long supported on religious and scientific grounds, dualist theories have been challenged by many variations of deterministic or reductionist models, in which consciousness becomes reduced to the product of environmental, genetic, or physiochemical factors. Behavioral psychologists such as B. F. Skinner believe that environmental factors determine human thought and action. Under these theories, the brain becomes, in effect, a passive conveyer of information, that is, an empty organism. In contrast, the genetic model assumes that behavior can be explained in the absence of conscious acts either by genes attempting to survive or by altruistic organisms operating instinctively by kin selection. E. O. Wilson, for example, concludes, "The brain exists because it promotes the survival and multiplication of genes that direct its assembly. The human mind is a device for survival and reproduction and reason is just one of its techniques" (1978: 2).

Although neuroscience research might ultimately validate or undermine the environmental and genetic models, its most direct relevance is to the physiochemical model. Despite the fact that Crick admits that the "brain is clearly so complex that the chances of being able to predict its behavior solely from the study of its parts is too remote to consider" (1979: 222), elsewhere he concludes that the "ultimate aim of the modern movement in biology is in fact to explain *all* biology in terms of physics and chemistry" (1966: 10).

Other observers contend that, at their base, mental processes are physical events. Patricia Churchland argues that a mental phenomenon might be "reducible to neurobiological phenomenon" (1986: 273). For Changeux, the combinational possibilities provided by the number and diversity of neural connections in the brain are more than sufficient to account for human mental capabilities. As a result, he feels that "[t]here is no justification for a split between mental and neuronal activity. . . . It seems quite legitimate to consider that mental states and physiological or psychological states of the brain are identical" (1997: 275). Similarly, according to the "Wet Mind" approach, the mind is simply what the brain does. "A description of mental events is a description of brain function and facts about the brain are needed to characterize these events" (Kosslyn and Koenig, 1992: 4).

Researchers in artificial intelligence and cognitive science reject the independent reality of the mind and envisage mental functions in a form that demands physical implementation. This functionalist approach focuses on what the mind does, for example, computing. Others like Paul Churchland and Daniel Dennett have said, in effect, that brains are to minds as computers are to processing. Under this model, according to Churchland, what we need is an "entirely new kinematics and dynamics with which to understand human cognitive activity, one drawn perhaps from computational neuroscience and connectionist A.I." (Miller, 1992: 180).

Critics of these approaches question whether the brain is complex enough to account for the mysteries of human imagination, mood, and memory. They find the reductionistic characteristics of any neuroscience explanation of the mind unpalpable and argue that consciousness will not yield to analysis. "Feelings, memories and experience are mental things that seem to be left out by neurophysical brain-talk, however detailed" ("Science Does It with Feeling," 1996: 75). While neuroscientists can explain in remarkable detail how the brain controls our body, they are far from explaining consciousness or demonstrating that there is nothing to the mind but the brain and its various physical states. This is not to say, however, that neuroscience will never be able to do so.

Consciousness

Perhaps the most mysterious aspect of the mind is consciousness or self-awareness, which can take many forms—from experiencing pain to planning for the future. Often the mind has been equated with consciousness, that which makes a human a human. Consciousness provides us with the continuity of our selfhood across our life. We are not only conscious of things but also conscious about our feelings about those things. We can even be conscious of our own feeling of being conscious about something (regress). Harth terms consciousness the "most challenging phenomenon exhibited by the brain" (1993: 133). But what is consciousness: a state of the mind or the activity of neurons?

As has research in other areas, neuroscience research has undermined traditional ideas about the unity or indissolubility of our mental life. Consciousness makes it appear that a single individual is the recipient of all sensations, perceptions, and feelings, and the originator of all thoughts. But according to Daniel Dennett and others, this apparent unity of the I and its self-awareness is largely an illusion. For Erich Harth, "[t]here is in the brain no single stage on which the multiple events picked up by our senses are displayed together" (1993: 133). Rather, consciousness is a process, a kind of global regulatory system dealing with mental objects and computations using those objects.

Interestingly, most operations of the brain take place outside our conscious awareness. They instead are carried out by a combination of genetic instructions and learned reactions to sensory inputs. We remain unconscious of most of our brain's activity. In fact, full awareness would be an impediment to our functioning. Restak (1994a: 129) notes that while the unconscious brain comes closest to a materialist's image of an intricate, thoroughly deterministic machine, the conscious brain is very different. At the highest levels of consciousness we experience a self-conscious controller who wills, remembers, decides, and feels.

Although many commentators equate consciousness with awareness, Restak (1994a: 130) disagrees. Though consciousness implies awareness, we can exhibit some lower levels of awareness without being fully conscious. However, because consciousness requires a vivid awareness of oneself as the experiencer, a prerequisite to consciousness is some level of brain activity and sensory awareness. According to Restak, "Consciousness must be understood as a very special 'emergent' property of the human brain. It is not an indispensable quality, since as we have seen the vast majority of the brain's activities do not involve consciousness" (1994a: 135). Although not all mental activities are accompanied by consciousness, then, consciousness cannot take place without such activities.

Though there appears to be little argument with the assumption that consciousness requires brain activity, there remains disagreement as to whether we can ever explain consciousness solely by the workings of the brain. According to Scott, for instance, consciousness is a real, "awesomely complex phenomenon" that cannot be reduced to some fundamental theory or a simple biological or chemical reaction (1995: 159). Similarly, while Churchland concludes that the state of consciousness is primarily a biological phenomenon, the contents of consciousness are "profoundly influenced" by the social environment (1995: 269).

Churchland's distinction between the state of consciousness and its contents appears to be supported by neurological evidence. It has been discovered that while the content of consciousness, as with memory, is found in the cerebral cortex, the maintenance and regulation of a conscious state is centered in the reticular formation region of the midbrain, which serves as an activation system for wakefulness. Because consciousness cannot occur without wakefulness, it is dependent on the activity of one of the most primal parts of the brain. However, since consciousness also requires content and a relationship to that content, it is always the product of interrelated activity of the neocortex and the reticular activating system, thus again manifesting the modular brain in action (Restak, 1994a: 126).

Changeux offers a theory of consciousness based on what he terms "mental objects" that allow us to be conscious of an "unending dialogue" with both the external world and the inner world of the self. The brain contains representations of the outside world in the neocortex and is capable of building representations of its own and using them in its computations. Because mental objects imply a much higher level of organization than that of a nerve cell, "the mental object is identified as the physical state created by correlated, transient activity, both electrical and chemical, in a large population or 'assembly' of neurons in several specific cortical areas" (Changeux, 1997: 137).

The assemblies noted by Changeux consist of overlapping sets of neurons possessing different singularities. Each assembly is discrete, autonomous, and closed, but not homogenous. The associative property of mental objects allows them to be linked together by sharing neurons. Moreover, a neuron takes part simultaneously in different mental objects while conserving its own singularities that existed prior to formation of the mental objects. Determining whether this theory best describes consciousness is problematic, but Changeux's conception of mental objects is valuable.

Visual Awareness

Until recently, the concept of consciousness proved too ill-defined to be explained quantitatively, thus leaving it to the realm of supposition. This situation changed when researchers, led by Crick and Koch, suggested that consciousness could be studied through neurobiological techniques if it was broken down. One elementary form of consciousness is awareness of one's surroundings and sensations, and of one's relationships to these stimuli (Barinaga, 1997: 1583).

Of all areas of awareness, the most substantial work has been concentrated on visual awareness. To understand visual awareness, it is important to ascertain how visual pathways come to represent what we see and how the nervous system influences these perceptions. Experiments have discovered that we do not perceive a visual image on the retina, but rather a neural image formed in the cortex. Although there are contributions from multiple processing systems in the neocortex, it is assumed that at some level there is an attention-directing command function that instructs the brain what stimuli to pay attention to. There is some evidence that this function may originate in the planning areas of the prefrontal cortex, also found to be important for memory.

People with normal vision regularly respond to visual signals without being fully aware of them. The brain uses past experience to help interpret information coming into the eyes. According to Crick and Koch, seeing is a constructive process in which the brain carries out complex computations. These computations involve the brain's actions in forming a symbolic representation of the visual world with a mapping of certain aspects of that world into elements in the brain (Crick and Koch, 1992: 154). Although the computations themselves are largely unconscious, we are conscious of the result of the computations.

According to the work in visual awareness, consciousness involves both attention and short-term memory (and possibly long-term memory). It would be impossible for a person to be conscious if they had no memory of what just happened. Iconic memory, which lasts only a fraction of a second, and working memory, which lasts a few seconds unless rehearsed, are both likely essential to awareness.

Given the interrelationship between visual awareness and memory, it is not surprising that although many regions of the brain are involved, the cerebral cortex plays a dominant role. Also critical are the corpus callosum, which connects visual information from both hemispheres, and the hippocampus, which is involved in short-term memory. The cerebral cortex, according to Crick and Koch (1992), draws on visual and other experiences to constantly rewire itself and create new categories. These categories, or mental objects for Changeux, are used in response to incoming visual signals to find those neural assemblies that, based on past experience, will best represent the objects or events coming through the senses.

Consciousness and Dualism

This work in visual awareness, if extended to consciousness in general, reinforces the view that consciousness is a unique property of the brain that is effectuated by a large number of interacting neural assemblies operating in parallel. Consciousness is inextricably tied to memory, reliant on attention-activating functions, and interconnected with sensory regions of the brain. Although we cannot substitute a description of physical brain events for consciousness, it does arise only through the joint activity of billions of neurons organized in assemblies or mental objects. For Restak, this evidence demonstrates that "[c]onsciousness, thought, memory, will, emotion—none of these has any independent outside reality other than in the context of the human brain. All are based on the brain's organization" (1994a: 13).

Before the death knell is sounded for dualism, however, it is important to note that there has not been a decisive resolution of the mind/brain question. Even if the mind is the expression of the activity of the brain and the two are in actuality inseparable, this does not mean that it is useless to separate them for analytical purposes. Though mental phenomena arise from the brain, mental experience also affects the brain, as demonstrated by many examples of environmental influences on brain plasticity (Andreasen, 1997: 1586). Scott concludes that it is not necessary to choose between materialism and dualism. Both can be accepted with certain reservations. Scott asserts, "We must construct consciousness from the relevant physics *and* biochemistry *and* electrophysiology *and* neuronal assemblies *and* cultural configurations *and* mental states that science cannot yet explain" (1995: 159–60).

The more neuroscience explains how the brain works, the more difficult will be the task of the dualists who demand an immaterial mind. With the rapid developments in our understanding of the mechanics of the brain, themselves products of the imagination of human minds, consciousness will lose some of the mystery that has surrounded it since at least the time of Plato. Although this is viewed as a threat by those who believe that we lose something special and private when we debunk the idea of the mind as separate from the brain, their fear of this shift to a modified materialism is probably premature because no matter how much we advance in neuroscience, it is unlikely the debate will disappear. Similarly, despite activities in artificial intelligence, information theory, and cognitive science that would reduce the mind to the workings of the computer, it is improbable that the mysteries of the human mind will be explained or replicated by even the most sophisticated computers imagined by the minds of humans. In the words of Jonathan Miller, "Consciousness may be implemented by neurobiological processes—how else?—but the language of neurobiology does not and cannot convey what it's *like* to be conscious" (1992: 180). The philosophical debate surrounding the mind/brain relationship and human consciousness will not abate, in spite of growing evidence of the importance of physiochemical factors for behavior.

POLICY IMPLICATIONS

What difference does it really make whether the holistic, localized, or modular theory of the brain is the most accurate? From a policy standpoint, as opposed to a philosophical one, does it matter whether the brain and mind are indistinguishable or separate? Although it would seem that none of this should matter, the resilience of the controversies over these topics demonstrates that they have significant policy ramifications.

First, throughout Western history, research on the brain has confronted opposition from both the left and right ends of the ideological spectrum. On the right, this research threatens the concept of the immaterial soul that is at the foundation of much religious doctrine. Even in a nonreligious context, the realization that the I is simply a product of a vast network of nerve cell connections challenges strongly held beliefs in individual responsibility, autonomy, and free will (see Chapter 5). Western theories of justice and their application in public law must be reevaluated in light of neuroscience findings. Moreover, such research threatens to take some of the mystery out of human life.

From the opposite ideological direction, the findings of brain research have always raised fears of the social impact of new discoveries: of social and behavioral control. The demise of the tabula rasa notion of the brain also has implications for education, social, and legal policies. Critics see threats of a psychochemical determinism that they, wrongly I believe, equate with past theories of genetic determinism. Brain research has also elicited opposition from liberals who fear that its findings will be used as weapons of oppression.

A second, more pragmatic, reason why these questions matter is that they have much to do with shaping the direction of research goals and funding. Research priorities follow accepted theories of science. The frictions between cognitive scientists and neurobiologists—and, more specifically, between those who would substitute computer chips for neurons and those who study neurons—are evidence of the competitive nature of brain research. In the end, how we view the brain can be self-fulfilling if the research is skewed too far in any one direction.

A third reason why these questions matter for policy is that the way we view the brain will have significant implications for treatment strategies. Are mental diseases disorders of a spiritual or cultural mind, or are they primarily biochemical imbalances or neurological deficiencies? Can localized treatment regimes be effective, or does a modular brain require multifaceted approaches? Is informed consent a meaningful concept in light of our knowledge regarding free will and autonomy? All these practical questions are addressed within our accepted standard of what the brain is, how it functions, and what its relationship to the mind is. Chapter 4 adds a complication to this equation—genetics.

4

Genetics and the Brain

One of the most promising and dynamic areas of research into the brain, but one fraught with many ethical and policy concerns and significant risks, centers on the interaction between genetics and the brain. Although such research is not nearly as advanced in terms of potential clinical applications as research in many other areas of neuroscience, it promises considerably more potential and wider applications both as basic science and as treatment. Moreover, should many of the initiatives on gene therapy in the brain succeed, the need for therapies such as neural-tissue grafting are likely to be obviated. As cautioned by the OTA, however,

> The brain, with its hundred billion or more cells and a thousandfold greater number of connections, seems likely to provide a virtually endless challenge for molecular scientists. Many people, however, may find that their discoveries, or even their hypotheses and research, are unsettling and disturbing. They bring into question familiar assumptions about human nature, responsibility and freedom, and basic equality among people (1988: 35).

The interaction between genetic research and our increased understanding of the brain promises to be a most critical political challenge to prevailing notions of humanhood and will perhaps redefine the debate over equality and inequality of humans. On a more practical level, this research sharpens the biopolicy issues inherent in the human genome project, particularly issues involving informed consent, human experimentation, safety, efficacy, and resource allocation.

THE GENETICS/BRAIN CONNECTION

Given the complexity of the workings of the brain as compared to the human genome, it is clear that no simple one-to-one relationship exists between them. It is estimated that the human genome contains approximately 100,000 genes, many of which are common to many species. Even the differential expression of all 100,000 genes, however, would fail to explain the extreme diversity of neuronal connections and the vast range of human behavior. Nevertheless, this fact does not negate the significant role that genes play in determining the boundaries and framework of the functioning brain. As noted by Changeux, a relatively small number of genes is sufficient to control the division, migration, and differentiation of the neurons shaping the neocortex. As illustrated in the brain-mediated model in Chapter 1 and the discussion of brain development in Chapter 2, the genes prescribe a template for neural functioning, but this template is completed by the environment and experience of each individual.

In addition to prescribing the generic template for the human species, genes provide the foundations for variation among individuals in terms of neural configuration and capacity. This impact is most obvious when dysfunctions occur due to deleterious genes or chromosomal abnormalities, as found, for instance, in Tay-Sachs disease or a fragile X chromosome. Discoveries from the human genome project (HGP) are positing many such direct linkages between the genes and the brain, and these findings are likely to accelerate in the coming years.

Many potential linkages between genes and behavior are already the focus of considerable controversy, such as the genetic bases of addiction, aggression, and risk-taking personalities. Whatever findings emerge, however, ultimately the power of the genes will not be sufficient to explain the details of neuronal organization, the precise form of every nerve cell, and the exact number and geometry of the synapses of any individual brain. If, however, the differential expression of genes is incapable of explaining the diversity and specificity of an individual's neural connections, what is?

One intriguing theory of the gene/brain linkage has been offered by Changeux. His epigenetic theory of selective stabilization is consistent with current knowledge of neuronal development and with our understanding of human genetic variability. Changeux contends that this epigenetic process does not require a modification of the genetic material because it acts not on a single cell but rather on a higher level of groups of nerve cells. He argues that the genetic "envelope" opens to more individual variability as we move up the evolutionary chain to humans. Whereas in other animals most behavior is genetically programmed, in humans it is not, thus opening human behavior to other influences.

The theory of selective stabilization assumes that the genetic influence is critical up to the point where the number of neurons peaks, soon after birth. (Although I would argue that this should not negate the potential environmental forces that can impact on neural development in the womb.) It is here that Changeux's model reverses what would be a more intuitive neuronal building process. What follows this point for Changeux is a growth process based on regression, as some neurons in each category die due to redundancy and some of the terminal branches or axons and dendrites of surviving cells degenerate.

Changeux uses language acquisition and hemispheric lateralization to support his theory. Language learning is accompanied by a loss of perceptual capacity, by an attrition of spontaneous sounds and syllables (1997: 244). Likewise, he argues that at a certain critical moment, similar if not identical neuronal structures exist in both hemispheres, but that they are lost selectively on the right or left early in childhood. For Changeux, "The word 'growth' should thus be understood in the sense of the lengthening and branching of nerve fibers, which eventually connect the cell bodies to each other (and to their targets) after the cells are differentiated and in place" (1997: 212). Under this theory, to learn is to stabilize preexisting synaptic combinations and eliminate the surplus. Therefore, activity can only be effective if the neurons and their basic connections already exist before interaction with the outside world.

Whether the process of learning and growth is based on selective stabilization as argued by Changeux or on the basis of the gradual building of new neural connections throughout life, the role of genetics is not deterministic either in terms of specific neural connections or behavior. Despite these limits, genes do exert a powerful influence on the brain and they are critical to our understanding of how the brain works. Rapid advances in the knowledge of molecular biology and applications of direct relevance to the brain are likely to complement similar developments in neuroscience. Although the genome cannot explain all the intricacies of the brain, we cannot explain these intricacies without a better understanding of genetics.

GENETIC RESEARCH AND THE BRAIN

We are currently witnessing an explosion in our knowledge about the functioning of the brain. One of the most vital areas relates to neurotransmitters, the chemical message carriers between nerve cells. In the recent decade we have identified over fifty substances involved with neurotransmission and determined some of their functions. The major goals of basic research are to understand the role each neurotransmitter plays in the body's chemical system and to discover how neurotransmitters interrelate and are balanced against one another. Ultimately scientists hope to find ways of supplementing deficient neurotransmitters and blocking effects of neurotransmitters that exceed the brain's needs, thereby developing the capacity to restore proper chemical equilibrium to brain and body. Neurogenetics hopes to develop the capability to isolate and analyze specific genetic defects in the neurotransmitter system, thus merging these two areas of knowledge. Already, the majority of the genes necessary for the functions of dopamine, serotonin, and norepinephrine have been cloned and sequenced (Comings, 1996: 84)

Genetic Diagnosis

One focus of the human genome project (HGP) is to identify genes that prevent normal brain development or that produce progressive brain degeneration. Identification of genetic markers for an expanding array of the neurological disorders discussed earlier is continuing, although to date our ability to alter the genes remains only a goal. Given the rapid progress in both our understanding of genetic disease and of brain functioning, many treatment regimes will be forthcoming.

At present, two types of tests are being performed on the DNA of cells taken from blood samples. For conditions for which the responsible gene has been identified, the tests use a DNA probe, or labeled segment of DNA, that binds directly to the defective gene if it is present. For those conditions for which the specific responsible gene is not yet known, genetic markers are identified that are close enough to the gene to be inherited with it. These restriction fragment length polymorphisms (RFLPs) indicate the approximate chromosome location of an unknown gene. By using overlapping RFLPs related to a gene disorder, the actual gene can eventually be isolated.

These indirect marker tests are more expensive, complicated, and probabilistic than the DNA probes. They entail getting blood samples from family members to determine which markers are inherited with that gene in the family, and then calculating the statistical probability that the gene is present in the person being tested. Through the use of increasingly sophisticated cloning and sequencing techniques, however, tests are being developed to uncover these markers in individuals. These techniques will be used prenatally or neonatally, or to identify adults who are carriers.

Genetic markers or genes have already been identified for Huntington's disease, Duchenne muscular dystrophy, and myotonic muscular dystrophy. In 1997 researchers from the National Human Genome Research Institute identified a gene on chromosome 4 that is associated with susceptibility for Parkinson's disease. Three identified genes are known to be associated with the chemical and neuropathological syndrome of Alzheimer's disease. These include genes found on chromosomes 21 (APP, AD1) and 19 (APOE, AD2), and most recently S182 on chromosome 14 (Roses, 1995: 80). Furthermore, genes or genetic markers have been isolated for lipid-storage diseases including Tay-Sachs, Gaucher's, and Neimann-Pick. In each case a reduced or missing enzyme results in an excessive buildup of lipids. Genes for these enzymes have been identified, and specific mutations have been found.

In addition, research is now under way to identify genetic factors that might predispose a person to be alcoholic. Although no single gene or gene complement has yet been found, "researchers are accumulating a great deal of information associating genetic factors with alcohol abuse" (Barnes, 1988: 415). In 1991 the U.S. National Institute on Alcohol Abuse and Alcoholism launched a massive study on the genetics of alcoholism. The $25 million budget for the first five years was to provide funding for the first systematic, multilevel study of the subject (Holden, 1991). Similar research on cocaine abuse is in the preliminary stages. Research is also progressing on the genetics of aging, though to date it indicates only a minor role of heredity in life span (Finch and Tanzi, 1997).

Until now most tests for genetic-linked diseases have relied not on identifying the abnormal gene but rather on detecting abnormalities in the gene product, such as the detection of an abnormal protein coded for by the defective gene. Rapid advances in molecular probes, however, are altering the givens. *Direct* tests for single-gene disorders depend on isolation and identification of the disease-causing gene. Similarly, *predictive* tests for polygenic or multifactorial disorders are dependent on the

TABLE 4.1
Risk of Mental Disorders, by Percent

	To general population	To first-degree relative (parent, child, or sibling)
Schizophrenia	1.0	9.0–13.0
Bipolar disorder	0.8	4.0– 9.0
Major depression	4.9	5.9–18.4
Obsessive-compulsive disorder	2.6	25.0
Panic disorder	1.6	15.0–24.7

Source: Office of Technology Assessment, 1992: 113.

identification of genetic markers that are associated with heightened rates of occurrence of the disorder.

Although single genes may be found that make individuals susceptible to major mental disorders, it is more likely that the mechanisms are more complicated and that the causes are multifactorial. Data such as that in Table 4.1, however, indicate a significant genetic component to schizophrenia, mood disorders, and anxiety disorders. The probability of having schizophrenia, for instance, when both parents are affected by it is 46 percent, well higher than the 1 percent of the population that is schizophrenic. The pattern of risk in Table 4.1 is similar across these disorders. Already genes have been identified that are associated with anxiety-related traits (Lesch et al., 1996) and an excitable or novelty-seeking personality (Bower, 1996), which might provide leads for more complicated mental disorders.

Genetic Therapy and the Brain

A long-term goal of the HGP as it relates to the brain is global gene replacement therapy in the CNS. Like other areas of somatic gene therapy, this therapy is likely to make use of viral vectors to produce a stable expression of normal human proteins in deficient cells, thus directly treating certain inherited enzyme deficiencies and metabolic disorders as well as cancers (see Neuwelt et al., 1995). Again as with gene therapy in general, major problems center on the delivery of genetic material to the correct site and the expression of the recombinant genetic material in the target cells. Various methods of transfer of recombinant genes into neuronal cells include direct transfer via microinjection, chemical methods, fusion using liposomes, the introduction of the gene into the germ line to produce transgenic mice, and virally mediated transfer, which at this stage shows the most promise.

Given the complexity of the brain, the yet incomplete understanding of particular functions and interactions among its parts, and the nature of the brain/mind relationship, however, gene replacement therapy of the CNS is especially problematic. The viral vector system used will need to provide both long-term and nontoxic gene expression to the specific neurons and glial cells in a precise region. If this is not done, the foreign genes may be expressed in inappropriate cell types or by nonneuronal cells (Latchman, 1996: 2). As a result, the dangers inherent in experimentation in this area are significant.

Genetic intervention in the brain is also likely to accelerate attempts to find the genetic bases of mental or behavioral traits and the molecular basis of memory. Development and delivery of biochemical aids to enhance the functioning of the pathways of communication between the different regions of the brain and to increase the capacity of the brain to integrate information have considerable potential. For instance, the molecular genetics of development suggest that variation in the speed of individuals in their ability to search and find processes might be associated with variation in certain proteins of their synapses. If these proteins could be genetically enhanced, these functions could be maximized. Furthermore, researchers are identifying with increased precision a variety of neuropeptides, hormones that the brain releases that are known to modulate functions such as blood pressure but possibly mood or emotions as well. Again, gene therapy might enable treatment of deficiencies or excesses in hormone production.

Until recently, it had been virtually impossible to transfer genes into CNS cells, either to study or influence molecular processes, largely due to the physical barriers that protect the brain and to the nondividing nature of neuronal cells. The blood/brain barrier designed to protect the brain severely limits blood-borne delivery of proteins and other macromolecules to the CNS, thus requiring alternative methods to deliver biologically active compounds to highly localized regions of the CNS. One such delivery approach is ex vivo, whereby cells are genetically modified in culture and later physically transplanted or grafted into the appropriate region. Another approach is the use of viral and nonviral agents to transfer therapeutic genes into neural cells.

The most commonly used gene delivery system, retroviral vectors, requires target cells that divide so that the transgenes can integrate and be expressed in the target cell's DNA. However, because neurons in the adult brain do not divide, retroviruses at present are of limited use in direct gene therapy in the CNS, although they are used to genetically modify cells capable of division in culture for grafting (Sena-Esteves et al., 1996).

Although retroviral vectors used most widely for gene therapy elsewhere in the body may still have significant research and clinical applications in the future, novel vector systems have emerged that allow exploration of the molecular basis of neuronal function and mapping of neuronal connections (Latchman, 1996: 3), as well as offering the prospect of gene transfer of DNA into neurons and other brain cells. One approach utilizes adenovirus as a vector for DNA delivery. An advantage of adenovirus from a safety standpoint is that a live oral vaccination of this virus has been administered to over 10 million persons to protect against natural adenovirus infections with no major side effects (Horellou et al., 1996: 43). Moreover, adenovirus recombinants have been used safely in the treatment of patients with cystic fibrosis. Key advantages of adenovirus or associated adenovirus vectors for neural therapy is that they can be delivered in vivo; they can transduce a variety of differentiated cell types, including nondividing neurons; and expression on the target cell can be restricted to the transgene only (Lowenstein et al., 1996: 12). Two major uses include the treatment of neurodegenerative disorders and brain tumors.

One likely application of gene therapy utilizing adenovirus vectors is to slow or reverse neurodegenerative disorders such as Parkinson's and Alzheimer's diseases. Combined with neural research designed to explicate our understanding of neurotransmitter mechanisms underlying disease symptoms and our knowledge of the genetics of neurotransmitters, such applications hold much promise. It has been demonstrated that these diseases originate from the progressive destruction of certain categories of neurons, thus causing impairment of neurophysiological functions of particular sites. For instance, Alzheimer's primarily affects cholinergic neurons, Huntington's primarily GABAergic striatal neurons, and Parkinson's dopaminergic nigrostriatal neurons. Therapy strategies attempt to either inhibit the degenerative process or stimulate regeneration of affected neurons.

Despite promising findings, considerable research is necessary, especially in the fine-tuning of cell specificity and levels of transgene expression, before human therapeutic applications are possible. Degenerative diseases are difficult to treat because of their diffuse nature, affecting large brain areas throughout the entire CNS as they steadily progress. Also, for many diseases, like Parkinson's, there are at present no presymptomatic diagnosis methods available, thus treatment can be started only after the symptoms become apparent, often after the degeneration of cells has progressed. In order to use gene therapy strategies for neurodegenerative diseases, problems limiting long-term expression in the brain following administration of recombinant vectors must be resolved. If this is accomplished, it will open up development of gene therapy for such diseases for which long-term expression is essential.

Coffin and Latchman (1996) argue that disabled herpes simplex virus (HSV) might be an ideal candidate to treat neurodegenerative diseases because it has evolved a life cycle in which infection is specifically targeted to neuronal cells and because it is capable of producing a latent infection which can be maintained for the lifetime of the cell. Although disabled adenovirus vectors can express a transgene for up to sixty days, HSV vectors promise the possibility of transgene expression for years after a one-time administration. Also, because HSV is a structurally complex virus, it can more easily be disabled either by deleting essential genes, and thus require complementation for growth in culture, or by deleting nonessential genes, those not required for growth in culture but necessary for growth in neurons in vivo (Coffin and Latchman, 1996: 103). Still, there are significant concerns about the safety of utilizing HSV vectors at this time.

Although gene therapy promises broad potential uses in the treatment of neurodegenerative diseases, the current availability of viral vectors that express themselves in the short-term at high levels of expression promises more immediate applications in the treatment of brain and spinal cord tumors. Approximately 15,000 brain tumors and 4,000 spinal cord tumors are diagnosed annually in the United States, and approximately 11,000 of these individuals will die. Even with the best conventional therapy, including surgery, radiation, and chemotherapy, the average life expectancy following diagnosis of the most virulent primary brain tumors is less than one year. Moreover, unlike other organs, skin, or muscle, destruction of even a small portion of CNS tissue by tumor growth or surgery to treat it can result in devastating consequences, because unlike these other systems, CNS tissue lacks the ability to regenerate following treatment. Also:

Diffuse infiltration of tumor cells into surrounding tissues, limited penetration of chemotherapeutic agents through the blood-brain barrier, and the sensitivity of neural tissue to ionizing radiation further reduce the efficacy of conventional therapies. (Shine and Woo, 1996: 54)

In comparison, gene therapy seems especially well-suited for treating CNS tumors because these tumors (1) are localized and rarely spread beyond the single site; (2) can be destroyed with short-term expression of cytotoxic genes; (3) are life-threatening with short survival times; and (4) are often malignant and aggressive, making total surgical removal impossible. Plus, any immune system response generated against the viral vector can be beneficial itself to tumor-cell elimination. Also, because tumor cells are generally the only rapidly dividing cells in the CNS, gene therapy strategies that target only dividing cells are especially effective (Shine and Woo, 1996: 54). The selectivity of retroviruses for proliferating cells that make them inappropriate for treating neurodegenerative diseases makes retroviral-mediated gene transfer especially promising for CNS tumors because it cannot affect the nondividing neurons. For this reason, Sena-Esteves and associates conclude, "In the next few years, we can expect a 'landslide' of new therapeutic modalities for brain tumors based on retrovirus-mediated delivery of genes designed to inhibit growth and to kill the tumor cells" (1996: 168).

There are many strategies for gene therapy to treat cancers, including: (1) the introduction of genes that alter the immunogenicity of tumor cells; (2) the introduction of genes that block expression of the oncogenes; (3) the insertion of tumor-suppression genes; and (4) the insertion of toxic genes into tumor cells. Another strategy that might be particularly effective for CNS tumors specifically is the introduction of genes that render tumor cells vulnerable to toxic agents.

In 1992, for instance, the National Institutes of Health Recombinant DNA Advisory Committee (RAC) approved an experimental protocol for transferring a viral gene into brain-tumor cells which made the cells susceptible to destruction by an antiviral drug (Stone, 1992). In the approved protocol, a retroviral vector carrying a gene from herpes simplex was inserted into mouse cells. The HSV gene codes for the enzyme thymidine kinase, which turns any cell producing it into a target for antiviral drugs. The mouse cells carrying this vector were then injected into the tumor through stereotaxic procedure, guided by magnetic resonance imaging. If successful, the retroviral vector infects nearby tumor cells, which then produce thymidine kinase, laying them open to attack through treatment by the antiviral drug ganciclovir. Animal studies showed that ganciclovir also killed other tumor cells in the area. The risks of such procedures are significant, and Shine and Woo caution that even if the first trials show promise, many refinements will be made before this procedure becomes a standard therapeutic strategy (1996: 67).

In combination with techniques designed to graft genetically engineered cells into targeted areas within the brain (Fischer and Gage, 1993) and direct intracranial administration of neuronal growth factors (Seiger et al., 1993), the use of viral vectors to deliver therapeutic genes offers a range of revolutionary techniques for intervention in the CNS. Gene therapy, then, is emerging as a potent new method for treating neurological diseases by introducing novel genetic material into the CNS of patients. In the future it might also be possible to treat other brain disorders such as addiction and mental illnesses through gene therapy, as well as to provide enhancement therapy and control eating disorders through genetic engineering of neurotransmitter functioning.

Although such interventions are not imminent and some might never eventuate, our rapid expansion of capabilities in combining the fruits of molecular biology and neuroscience promises many challenges. As we develop second- and third-generation vectors designed with less toxicity to normal tissues, increased expression of the recombinant gene, and improved specificity, our capacity to intervene will extend beyond anything now imaginable.

POLICY ISSUES IN GENETICS AND THE BRAIN

This move from diagnosis to therapy raises many policy issues regarding what role the government ought to play in encouraging or discouraging such research and application. It also raises ethical questions concerning parental responsibilities to children, societal perceptions of children, the distribution of social benefits, and the definition of what it means to be a human being.

In light of new genetic interventions in the brain, questions arise as to what constitutes a deficiency or disorder. This question is particularly poignant as linkages are found between protein levels and characteristics such as personality traits (Cloninger et al., 1996), sexual orientation, and aggression. As noted earlier, researchers have identified a gene (D4 dopamine receptor gene, or D4DR) linked to a novelty-seeking or excitable personality (Bower, 1996), while another variant has been linked to neuroticism (Lesch et al., 1996). Moreover, though gene therapy is now focused on identifying specific genetic factors in neurological diseases or disorders, pressures for gene enhancement are likely to follow, particularly in the United States, with its competitive culture and faith in technological fixes. As will be discussed later, like eugenics, the history of intervention in the brain is a controversial one. This controversy will intensify as associations are found between genetics and brain function, mental disorders, addictive behaviors, and social deviances.

Neuroscience-based research is central to the recent advances in our understanding of addictive behavior, ranging from how genetic and environmental factors relate to addiction, to how addictive substances act on the brain and how substance abuse might best be treated. Also, according to the Institute of Medicine, the challenge is to come to terms with human aggression and the "destructive strain that humankind seems to carry from one generation to the next like an inherited disease" (1992: 11). Neuroscience, in combination with molecular biology, has the means to provide such a base of knowledge. If a biochemical explanation can be found for violent and aggressive behavior, such knowledge could lead to medications for treatment or to efforts at behavioral control. The line is a gray one and depends on how the condition is defined by the medical community and society.

The faces of inequality therefore are likely to change as more knowledge about the genetic bases of brain activity and functioning accumulates and as our capacity for intervention expands. Questions about the genetic bases of criminality have already engendered intense controversy (see Masters, 1996). Although most observers agree that it is unlikely that genetics will prove to have proximate linkages to criminal behavior, even the notion of genetically based criminal tendencies poses severe constitutional issues. Experience with XYY testing and research in the late 1970s demonstrates the explosiveness of any research into the genetic/brain ties with criminality or with any form of antisocial behavior.

Another application of knowledge of genetics and the brain on perceptions of inequality is found in recent revelations regarding homosexuality. Studies that found that the area of hypothalamus believed to control sexual activity was less than half the size in gay men than in heterosexual men, combined with announcements of the identification of a "gay" gene, were met with strong and devisive reactions. How this information was received depended primarily on the perceived political context. On the one hand, some gays welcomed the news on the grounds that it demonstrated that homosexuality is natural, not learned. If homosexuality is genetic and controlled by the brain, then civil rights protection based on the notion of immutable characteristics might be warranted (see LeVay,

1994). On the other hand, in a homophobic society, the search for genes for susceptibility to being homosexual could lead to prenatal screening or genetic therapy designed to eliminate persons with the "aberration" or to treat them for it (see Schuklenk et al., 1997).

Research into genetics and the brain, then, promises to accentuate the already acrimonious political debate over human nature, personal identity, and equality. Traditionally, differences both in genetic complement and behavior have tended to be defined as diseases, disorders, and conditions to be treated. The way we as a society respond to these remarkable technological advances and the knowledge that spawns them depends to a large extent therefore on our conceptions of equality and inequality. There is historical evidence to suggest that the forces that embrace such findings as proof of inequality are strong. Commitment to the view that all humans are innately equal will face powerful challenges in ensuring that new knowledge is directed toward that end. This will be reflected at two levels: (1) Should society pursue certain areas of research, and (2) under what conditions should individuals be encouraged or required to undergo the types of treatment being developed?

Eugenics and Genetic/Brain Intervention

Although the term eugenics is attributed to Francis Galton, who coined it in 1883 to refer to positive attempts to improve the human gene pool, eugenics as a concept was fundamental to Plato's republic and has been practiced in various forms throughout history. Eugenics in its most base form was practiced by the Nazis, without the more sophisticated technologies of the new genetics. How eugenics is practiced is a function of the state of technology and is dependent upon the forms of genetic intervention that are feasible. Conversely, the concept of eugenics at times has been instrumental in shaping what is claimed to be scientific fact.

Eugenics is now becoming an intense critical force in debates over the direction of science and the application of specific types of genetic/ brain intervention. Furthermore, the debate in the future is likely to be less clearly defined and more complex than the debate that took place in the United States in the pre–World War II period, due to the fact that it is considerably more difficult to identify the motivations behind particular policy proposals.

Because of the extremely negative connotation attached to the term eugenics, only the most adamant proponents of state-controlled human procreation use explicit eugenic arguments (see Lykken, 1997). Instead, most proponents of intervention policies couch their stands under less pejorative justifications such as:

- reducing social or family burdens
- benefits to future generations
- social responsibility
- cost containment
- individual health benefits
- individual rights to a sound mind and body

Distinctions must be made therefore between explicit forms of eugenics, such as compulsory sterilization policies, and less innocuous forms, such as preimplantantion or prenatal diagnosis for Down's syndrome or screening for predispositions for schizophrenia, alcoholism, or homosexuality.

One important dimension of the eugenics debate is currently focused on the question of who makes a eugenic decision. At the center of the most infamous eugenic programs of the past has been a presumption of compulsory implementation under the auspices of the government. Current and future genetics policies in the United States, however, are more likely to be designed to minimize at least the appearance of compulsion. In fact, compliance will most likely result from social pressures that define responsible behavior—a consequence of a causal logic that implies persons have a duty to use available technologies that are deemed socially appropriate. According to the Office of Technology Assessment, "new technologies for identifying traits and altering genes make it possible for eugenic goals to be achieved through technological as opposed to social control" (1988: 84).

Annas (1989: 20), for instance, questions whether the routinization of preimplantation screening of embryos constitutes eugenics if, rather than being required to use these services, parents are propagandized to want and even to insist upon screening as their right. Does the voluntary but widespread practice by a population in using genetic intervention to increase the number of desired traits in offspring constitute eugenics? If not, what level of government involvement (incentives, education, provision of services, economic coercion, compulsion) is necessary for practices to be considered eugenic? What happens if commercial firms are able to convince the public through effective mass marketing of their products that such services are in their best interests? Is there any way we can use genetic technologies to raise the norm of intellectual capacities without being accused of practicing eugenics?

One observer sees the form of eugenics now "creeping back into science" as different in detail but identical at its base to the eugenics of the mid-twentieth century:

> But like the endless number of movie sequels that have overrun our movie theaters, eugenics is back with a new cast of characters and a slightly different script, but the same tired and dangerous plot. (Allen, 1989: 9)

Allen argues (1989: 11) that once again the proponents of eugenics are using highly speculative research on biological diversity as the basis for social programs to turn attention away from environmental causes of social behavior, such as wage and benefit cuts, inflation, and stress in the workplace. For observers of Allen's persuasion, the new forms eugenics is taking are more insidious than the old forms because policies are being implemented under the guise of the socially acceptable goals noted earlier. Due to the lack of a consistent definition of eugenics, however, any genetic intervention, no matter how benign, is open to the charge that it is "eugenic" and, therefore, by association with past eugenic policies, should be prohibited.

The reemergence of a concern for human genetics in the 1980s was accompanied by a shift in emphasis back to the positive eugenics of Galton. The discredited negative eugenics approach, with its dependence on compulsory sterilization and marriage restrictions, is currently being replaced by a subtle but far-reaching trend toward approaches designed to maximize propagation of the most genetically fit. Largely, this shift in emphasis is the result of an array of new reproductive technologies. Although cryopreservation of human semen, ova, and embryos in combination with sophisticated fertilization and transfer techniques will allow people with genetic problems to have children, they also provide considerable opportunity to practice positive human breeding. As advances in in vitro fertilization with embryo screening for chromosomal abnormalities continue, the focus of future eugenics efforts could shift from adults to preimplanted embryos unless caution is exercised in their use.

More important to social policy than the new technologies are the changes in social values that lead to their widespread application. Prenatal diagnostic techniques now enjoy substantial support among the medical profession and the general public. As new diagnostic tests and genetic probes emerge, public expectations are likely to intensify and the demand for accessibility to information derived from such efforts will heighten.

Once the tests become accepted by policy makers as legitimate, it is likely that legislatures and courts will recognize professional standards of care that incorporate these tests. Although genetic probes cannot account for all the phenotypic variance in the expression of the genes, many people are likely to perceive a positive gene probe test as an indicator of a person's biological destiny. Lappé contends that should this tendency prove true, DNA probes might "acquire a misleading status in our medical armamentarium as indicators of a new kind of biological determinism" (1987: 10).

To date, genetic technologies have targeted diagnosis and, in a few cases, amelioration of genetic anomalies, not alteration of human capacities through positive eugenic manipulations. But this situation is changing. Research surrounding the fragile X chromosome, although now focused on aberrations, might eventually provide the means to expand mental development significantly. Findings of the genetic bases for alcoholism, personality development, memory, and longevity introduce similarly critical implications for a new eugenic thrust in science. Increasingly, then, the unfolding knowledge of the human genome promises attempts in the near future at "improving" human nature. As the mechanisms by which the genes influence behavioral characteristics are discovered in the coming decades, however, the boundaries of the nature/nurture debate are likely to change drastically and intensify. As stated by Lappé, "As a minimum, these events suggest that we will be uncovering knowledge for which we are unprepared, and some that will challenge tightly held convictions about the predominance of nurture over nature" (1987: 5).

It is not surprising that a cultural mentality that urges us to use all available technologies to maximize propagation of a genetically strong species would be attractive in a society that is oriented toward being a leader in all endeavors. Cultural values that supported eugenic sterilization and marriage laws are even more amenable to the acceptance of new genetic techniques. With the potential availability of techniques to genetically engineer traits that convey a socially competitive advantage for individuals, parental demand to maximize their progeny's chances will be substantial. Moreover, those parents who are unable or unwilling to use these technologies might find their offspring condemned to a second-class citizenry, while that which had once been considered within the range of the normal gradually slips into the domain of the subaverage. Given our litigious society, it would not be surprising to see children sue their parents for failing to use available genetic enhancing technologies. Torts for "wrongful life" might be extended to include such inaction on the part of the parents.

This desire for children who are "perfect" has significant long-term implications for our perceptions of children and childhood. As couples and singles limit their families to one or two children, the demand will increase for technologies which can guarantee the "perfect" child. According to many observers, young couples are now going to considerable lengths to ensure the birth of a near-perfect child, including sex preselection. As more precise and effective genetic techniques emerge, pressures for access will intensify. Survey data indicate that many parents would consider termination of a pregnancy for even moderate defects in the unborn child, such as a heightened risk to early heart disease or criminal tendency. For instance, one survey (Keeton and Baskin, 1985) found that 76 percent of the respondents would consider an abortion if their fetus was diagnosed as having any one in a range of serious disorders. Thirty-eight percent would consider an abortion of a fetus identified as having a heightened risk of early heart disease, muscular dystrophy, or blindness. If possible, 11 percent of the respondents would consider making genetic changes in intelligence, and 83 percent would consider the use of gene therapy for health reasons.

A couple that does not make use of routine screening techniques and produces a child with a technically avoidable congenital defect under this value system might be accused of acting irresponsibly and imposing a burden on society. Even if parents exercise free choice by rejecting the use of potential technologies, some elements in society might look only at the results of that choice and, based on societal norms, judge the action to be irresponsible. This obviously is a dangerous "causal logic," especially for the target groups, that is, those who somehow escape the technological screening mechanisms. This scenario also demonstrates that what on the surface might look like a voluntary and positive means of enhancing the genetic complement of one's progeny can easily produce underlying and intense societal pressures, leading to coercive state intervention.

CONCLUSIONS

The interface between genetic and neuroscience research is one of the most fascinating and dramatic areas of biomedical research. The possibilities for using the fruits of molecular biology to treat and possibly avert neurological disorders appear tremendous. In combination with an array of assisted reproductive techniques and preimplantation and prenatal diagnostic procedures, there is considerable hope for improvement of the health of our progeny. As pointed out here, however, these presumed benefits carry with them risks, not only to individual patients or subjects but also to future generations.

Rigorous social research is necessary to analyze the value changes that accompany this rapidly expanding arsenal of technological interventions emerging from genetic/brain research. Are the potentially threatening value shifts toward eugenics real or imagined? If there is substance to the fears expressed by critics, what policies can be implemented to minimize their materialization? What can be done to ensure that these technological advances are applied within a social environment that is not conducive to eugenic motivations? Furthermore, what policies, if any, are needed to protect patients and subjects involved in this research and guarantee their informed consent?

5

The Brain and Behavior

As demonstrated in Chapter 4, the genetic links to behavior are increasingly apparent, largely the result of genetic research under the human genome project. Although applications are yet limited, genetic diagnosis capabilities will provide us with an array of predictive tests, of varying reliability and preciseness, for susceptibilities to many personality and behavioral traits. Moreover, in the next decade gene therapy will likely emerge for many of these traits. As a result of this new genetic knowledge and its applications in testing, diagnosis, and therapy, the linkages between the genome and behavior are again becoming matters of heated controversy (see Parens, 1996).

Ultimately, the controversy over behavioral genetics will focus on the brain because, as posited in Chapter 1, it is through the neural system that genetic influences are manifested. Furthermore, the more we understand the functioning of the brain, the more we are led to the conclusion that we as individuals are limited by what our individual brains permit. This is not to be interpreted as suggesting that our brains determine behavior, but rather that they mediate genetic and environmental influences. In the words of Roger Masters:

> The neurochemistry of behavior is not the same theory as genetic determinism. On the contrary, neurotransmitters like serotonin vary from one individual to another for many reasons, including the individual's life experience, social status and diet. Genes may influence neurochemistry. So do behavior, culture, and the social environment. (1994: xiv)

Ultimately, however, the brain affects or mediates every action and thought of both political leaders and citizens. Our capacity for enjoyment, suffering, and behavior to some degree is inscribed in neurons and synapses. As a result, our interpretation of the world, including the political and social dimensions, and our responses to it are dependent on the internal organization of the brain. Therefore, in order to make sense of human behavior we, by necessity, must understand the organization and functioning of the system that controls or modulates it, the central nervous system.

More specifically, (reductionist) attention needs to be placed on neurotransmitters. Because they are the important communication links between neurons, they logically are the prime targets for explanations of behavior, especially when they malfunction. Those neurotransmitters which appear central to behavior are dopamine, which is crucial to the regulation of motor behavior; serotonin, which handles much of impulse transmission; norepinephrine, which is involved in transmitting impulses to the autonomic nervous system; and gamma aminobutyric acid (GABA), which inhibits neurons from firing or sending impulses. The understanding of these neurotransmitters (as well as epinephrine, glycine, glutamic acids, and the endorphins) is likely to provide a better appreciation of the role of the brain in human emotion and behavior.

The findings of neuroscience, then, require a model that acknowledges that the brain has a major role in explaining behavior. As noted by Changeux:

> The development of the neurosciences has brought another way of looking at behavior. . . . The neuronal content of the black box can no longer be ignored. On the contrary, all forms of behavior mobilize distinct sets of nerve cells, and it is at their level that the final explanation of behavior must be sought. (1997: 97)

Although it is debatable whether we will ever be able to describe a particular behavior in terms of specific neuronal activity (I doubt it), it is crucial that this neural dimension be an integral part of any respectable paradigm of behavior. To ignore the role of the brain is no longer possible in light of what we now know, even in the rather primitive stages of neuroscience.

Current behavioral science theory, especially, will have to be qualified by neuroscience findings. Although uncomfortable to many behavioral scientists, new knowledge of the brain must be incorporated if the behavioral sciences are to retain credibility as science. "Even now the new developments in the biochemistry of the brain and in psychopharmacology demand attention by social scientists, if only at the public policy level" (White, 1992: 16). Although inclusion of the biological sciences in the social sciences is in no way an original theme (see Wiegele, 1979; Blank, Caldwell, Wiegele, and Zilinskas, 1998), dramatic advances in the neurosciences over the past decade make it more imperative.

The rapid development of cognitive science along with behavioral genetics has already led to significant alterations in psychological theories of abnormal behavior. "Perhaps the greatest source of optimism and excitement in the field of abnormal psychology in the last twenty-five years has been the tremendous advance in the study of the biological bases of behavior" (Bootzin et al., 1993: 88). This enthusiasm, however, has not generally extended to the social sciences, particularly political science.

As critical as it is that a neuroscience dimension be incorporated into explanatory models of behavior, this does not mean that conventional factors should be neglected. In criticizing behavioral genetics, Parens notes that "as long as our society seeks simple explanations for phenomena as complex as the differences between individuals and groups, danger looms" (1996: 17). I would add that while this means that the biological bases of behavior, both genetic and neuronal, should not be viewed as deterministic forces, any model that minimizes their cumulative impact is not only simplistic but deceptive.

As discussed in Chapter 1, biological models of behavior, even when heavily moderated by environmental influences, will always be controversial in Western societies. They challenge the foundational concepts of democracy: equality, individual freedom, and free will. They also suggest that social change will not necessarily lead to desired changes in behavior. We shall now examine several selected areas in which neuroscience raises concern: the biological bases of violence; addictive behavior; sex-related differences; and sexual orientation. The chapter concludes by returning briefly to the topic of free will and individual responsibility.

THE BRAIN AND VIOLENCE

No credible scholar today would argue that the causes of violent or aggressive behavior are either all environmental or all biological. Although in exceptional cases these behaviors might be entirely biologically or environmentally based, in the overwhelming majority of cases they cannot be traced to any single factor (Comings, 1996: 84). A vast literature across many disciplines convincingly demonstrates that violent behavior even of a single person is the result of a combination of overlapping and often reinforcing forces. For Greenspan:

No controversy about the predominance of nature or nurture in human development should exist. A child's constitutional makeup interacts with his emotional experience in a reciprocal manner so complex that there is no point in debating which factor contributes more. (1997: 133–34)

The most appropriate approaches, therefore, are those that explicate how biology and environment are related—how a complex of biological factors interact with and influence a complex of environmental factors to produce a violent or aggressive behavior or behavioral pattern. A neurological perspective by itself, then, is not determinative of behavior, and in fact it is likely less explanatory than combined social, economic, and genetic factors. Also, because of these complexities, it is risky to generalize from individuals to groups in looking for biological ties to violent behavior.

It should also be noted that the line between biological and environmental factors is blurred. Many brain deficits that are related to violent behavior are themselves the result of environmental insults. The use of alcohol has long been known to provoke aggressive and violent behavior in some people, as have low-cholesterol diets, steroids, and drugs of abuse (Kotulak, 1996: 64). Moreover, environmental carcinogens, mutagens, and teratogens are capable of producing tumors or developmental injuries. Lead exposure to fetuses and children is especially risky for neural development. The full impact of workplace and other environmental neurotoxins is far from being recognized (see Chapter 8) and demonstrates the sensitive linkages among disparate influences on behavior.

With this more complicated context in mind, it is important to look at neuronal contributions to violent behavior. If we are to understand the complex interactions among the various contributors to such behavior, the role of the brain and its influence must be clarified. This section briefly examines what is now known about the brain's contribution to this equation. The absence of equal time to the social factors should not imply, however, that the neurons act in isolation.

The constellation of related behaviors, including violent, aggressive, criminal, antisocial, and impulsive, are often studied together despite their varied implications. Although there is no specific gene or neuronal pattern for any of these manifestations, there is considerable evidence of neuronal disorders that predispose children to impulsive, hyperactive, or aggressive behaviors that in some cases persist throughout life. Episodic dyscontrol, the result of seizures in the limbic system, for instance, is a well-documented disorder that can lead to abrupt, unexplained acts of rage. Violent outbursts, including aggressive behavior in automobiles, can be traced to this disorder. Studies have found that 94 percent of persons with uncontrollable rage have developmental or acquired brain defects (Restak, 1994a: 151).

Brain laterality, for instance, has been found to be associated with antisocial behavior, with some evidence of a higher incidence of left-handedness among criminals. Moreover, several studies found that 76 percent of violent offenders and 91 percent of the psychopaths studied had evidence of left-hemispheric dysfunction in the temporal and frontal lobes (see Jeffrey, 1994: 164). In addition to frontal and temporal lobe abnormalities, violent behavior tends to be correlated with abnormalities in the amygdala and other areas of the limbic system.

A very high rate of injuries and trauma to these areas of the brain has been found among criminals. These people's injuries can be the result of birth injuries, childhood illnesses, exposure to neurotoxins, accidents, or, ironically, violent acts to themselves. One study found that 70 percent of the violent offenders examined suffered from head injuries, and another study of fourteen juvenile death-row inmates concluded that all fourteen had brain trauma or neurological disorders. Moreover, thirteen came from families with a history of violence and twelve had been brutalized sexually and physically as children (Lewis, 1988), again indicating the social dimensions of neurological ties to violence. Supporting the interactive nature of environment and brain, hair analyses of serial murderers and violent offenders have demonstrated excessive concentrations of lead and cadmium in such individuals (Jeffrey, 1994: 165).

Frontal lobe injury has long been correlated with hyperactivity, impulsiveness, and aggressive behavior. Early EEG research found relationships between abnormal electrical discharges in the brain and behavioral problems, and through more sophisticated brain imaging these data are becoming more precise. Such studies demonstrate that abnormalities are identifiable in 15 to 50 percent of violent people, as compared to 5 to 20 percent of those persons with no history of violence. What this means, however, is not clear because the deficits are often not apparent in behavior. Also, the range of expression of similar brain injuries is enormous (Restak, 1994a: 152) since no two brains are identical. Furthermore, the brain is remarkably adaptable and may compensate for damage.

The focus of research on the neural influences on aggressive behavior centers on the neurotransmitter serotonin, although noradrenaline, norepinephrine, and dopamine have also been targeted. Serotonin was first implicated in research showing that people who became aggressive under the influence of alcohol had lower levels of this neurotransmitter than those who did not become aggressive. Although low serotonin levels do not compel a person to be violent, they appear to lower the threshold for violence (Kotulak, 1996: 88). Because serotonin normally acts as a brake on impulses, a deficiency means that the person, in effect, does not have full control. (For a comprehensive analysis of the role of serotonin in criminal behavior and its implications for law see Masters and McGuire, 1994.)

Hormonal levels have also long been linked with behavior. Carey and Gottesman, for example, argue that we have already found the genotype that predicts violence better than any gene to be discovered in the future: the XY genotype. As discussed later in this chapter, heightened levels of testosterone have a significant role in sexual differentiation and sex-typical behavior. Males on average are more aggressive and potentially violent than females, and again the brain plays a critical role in the regulation of hormone production.

Although the brain damage, hormonal, and neurotransmitter arguments have been presented as defenses against conviction for violent crimes, rarely have such attempts been successful (see Shapiro, 1994). In part, this is because the notion of brain damage or abnormality remains subjective and the links to any specific behavior are tenuous at best. We are far from understanding how the brain influences aggressive and violent behavior because each act has multiple influences and violence is such a diffuse concept. Due to its greater specificity, we are much further along in delineating the neural basis of addictive behavior.

THE BRAIN AND ADDICTIVE BEHAVIOR

Addiction is a major social and health problem in U.S. society that increasingly has been linked to the brain. The financial cost of alcohol abuse alone is estimated to be over $90 billion annually. Other substance abuse adds $70 billion to this cost. Over 30 million Americans (approximately one in six) alive today will experience addiction to alcohol or illegal substances in their lifetime. Moreover, 40 percent of American families are affected by addiction. One could argue that the $92 billion spent on alcohol, $44 billion on tobacco, and $40 billion on major drugs of abuse (cocaine, $17.5; heroin, $12.3; marijuana, $8.8) in 1990 alone could have been put to better use.

Furthermore, it is estimated that one-quarter of deaths in the United States are caused by the use of tobacco, alcohol, and illegal drugs.

According to DuPont, "Addiction is the number one preventable health problem in the United States and throughout the developed nations of the world" (1995: 4). Although the social and cultural dimensions of addiction are complex, attention here is focused on the relationship of addiction to the brain.

Two issues of addiction that are relevant here are: (1) the biochemical/genetic bases of addiction and (2) the impact of addictive substances on the brain and its normal functioning. Findings from neuroscience research in the past several decades have illuminated addictive behavior by explicating the role of genetic and environmental factors and their interaction with the biochemistry of the brain. Through expanded knowledge of the roles of specific neurotransmitters and the ability to visualize the brains of addicts through PET imaging, the neural bases of addiction are becoming clear. As our understanding increases, it is becoming obvious that addiction extends far beyond the physical need for chemicals to a wide range of activities (eating, gambling, sex) that produce feelings of dependency in our neural networks.

Although addiction can affect all organs of the body, the primary target is the brain (Nestler and Aghajanian, 1997: 58). Addictive substances or behaviors are linked to the brain's capacity to experience feelings of pleasure and pain, a capacity that has evolved to manage fundamental behaviors such as feeding, reproduction, and aggression. When the brain's pleasure centers are stimulated, the brain sends out signals to repeat the pleasure-producing behaviors. According to DuPont, the brain is selfish and characterized by the "right now" quest for pleasure. "When it comes to many natural pleasures, the brain has built-in protections. It has powerful feedback systems to say 'enough' when it comes to natural behaviors, including aggression, feeding, and sex" (1995: 5). The brain is selfish, however, in that automatic brain mechanisms do not account for delayed gratification. Therefore, when the brain comes into contact with an addicting substance and this substance triggers the pleasure centers, there is a strong incentive to repeat the exposure. These feelings, of course, are constantly mediated by culture and other environmental forces that can influence behavior.

To complicate matters further, there is strong evidence of genetic predispositions to addictive behavior, and possibly to addiction to particular substances such as alcohol. Moreover, people who are genetically oriented toward immediate gratification, or to impulsive behavior and risk taking, are also at higher risk for addiction. Despite the importance of genetic, cultural, and social factors in explaining addiction, at the base our understanding must focus on the brain (Leshner, 1997). Not only is the brain the key to unlocking the causes of addiction, but the brain must also be the focus of study to determine how addictive drugs and behaviors affect the functioning of the brain, cause distortions in thinking, and change the brain of the addicted person.

Although a wide range of behaviors are potentially addictive (e.g., sex, gambling, eating, running, and surfing the Net), most attention has been focused on drugs because it is with chemicals that the effects are most apparent. All drugs of abuse produce their effects by traveling through the bloodstream to the brain. Once in the brain, each drug alters the function of specific brain cells. Stimulants such as cocaine act as an exciting influence on certain nuclei, while depressants such as alcohol and narcotics act to inhibit activity of these nuclei. Some drugs act by blocking the reuptake of neurotransmitters from the synapse to the sending axon, thus facilitating transmissions by prolonging the time the neurotransmitter remains in the synapse. Other drugs actually mimic particular neurotransmitters by sending their own messages and occupying the receptors. Moreover, some substances like alcohol interfere with the cell membrane, while others affect the synapse, working either as agonists (activating transmission across the synapse) or antagonists (blocking the receptor sites on the dendrites).

Whatever the precise mechanism of a specific substance, tolerance for it builds. This happens because when a particular neurotransmitter is excessively stimulated over a long period of time, the brain reestablishes an equilibrium by reducing the sensitivity of the affected receptors or by decreasing their number. This process, termed down-regulation, means that the more the brain is exposed to chemicals affecting a neurotransmitter, the less the brain responds to that specific dose. Therefore, in order to experience the same effect, the addict must use increasingly higher doses. A related effect, physical dependence, is manifested by withdrawal symptoms experienced when use of the substance is stopped. Such symptoms vary by substance and reflect the cellular adaptation of the neurons of that area of the brain to the continued presence of the substance that has influenced their functioning. Withdrawal symptoms manifest the shock to the brain from a rapid alteration of the chemical environment. Frequently, these symptoms are interpreted by the addict as the "need" to resume use of the substance.

The key to understanding the biochemical bases of addiction, then, is at the molecular and cellular levels in the mechanisms of neurotransmitters (Nestler and Aghajanian, 1997). Two prominent theories of addiction focus on the endorphins and dopamine. According to the latter theory, most of the drugs of abuse, including alcohol, cocaine, amphetamines, and narcotics, stimulate the dopamine-producing neurons in the median forebrain bundle, the neural pathway that connects the midbrain to the forebrain (Wise, 1988). This increased production of dopamine creates the euphoria and pleasure associated with the high, thus reinforcing the substance's continued use. Research has demonstrated that if dopamine production is turned off by dopamine-suppressing chemicals, the stimulating effects of the drug are blocked.

The second theory, applied specifically to opiate addiction, focuses on a group of peptides, the endorphins of which more than a dozen natural forms are known. The endorphin brain system moderates pain, promotes pleasures, and manages stress. Endorphins also act as neurohormones and can affect nerve functioning at distant sites in the nervous system through the blood. Endorphin receptors are found in other parts of the body including the intestines, which might explain why these drugs often affect other organs as well.

It has been postulated that endorphins can explain the physiological dependence on heroin, because when external opiates are taken, the brain ceases to produce endorphins. As a result, the person becomes totally dependent on the drug for relief of pain or a feeling of pleasure since natural production by the brain of these needed chemicals has ceased. Termination of the drug use results in withdrawal symptoms until the brain resumes endorphin production (Bootzin et al., 1993: 324). Furthermore, research indicates that the opiate receptor sites can be occupied by antagonists such as naloxone and naltrexone, which are used to treat overdose and addiction. Even if the opiates get to the receptor sites first, the antagonists cover the sites, thereby blocking the drug's capacity to produce a rush.

Not surprisingly, the effect of a particular substance as well as its addictive properties depends on many factors, including chemical composition and purity; dosage, timing, and frequency of exposure; and the route of administration. Because the most rewarding drug experience is achieved when the brain is hit by a high and rapidly rising level of the chemical, injection directly into the vein is the most effective delivery route for most substances. In turn, smoking is more addictive than snorting or taking the same drug orally.

Returning to the two questions that framed this section, what scientific evidence is there that the root of human addictive behavior lies in the brain and what are the dysfunctional effects of addictive substances on the brain? Although the brain's role in addiction has long been a matter of speculation, research on the neural bases of addiction began with experiments on animals in the 1950s that utilized electrodes implanted in the pleasure and pain centers of the brain. More recently, knowledge of neurotransmitters and improved instrumentation allow for precise chemical probes of specific brain nuclei. The general finding of an extensive body of research from the 1970s and 1980s is that while the various substances act through a wide array of distinct mechanisms, ultimately they all work to stimulate pleasure centers and suppress pain centers (Leshner, 1997: 46). This commonality in result explains why addicts are willing to use diverse drugs in their search for a high.

Specifically, researchers have discovered that several areas of the brain—the ventral tegmental areas and the nucleus accumbens—exhibit high concentrations of dopamine–containing neurons and that all drugs of abuse trigger the release of relatively large amounts of dopamine into the synapses of these neurons, albeit through varied mechanisms. Critically, this research demonstrates that once introduced to the effects of these substances, these neurons require a repeat exposure to activate the release of dopamine and to produce the pleasurable response again, resulting in the reward pattern of addiction. As we come to better understand the function of neurotransmitters and receptor sites and the mechanisms by which drugs influence neural activity, we should be able to determine why some people are more susceptible to addiction than others and to offer preventive treatment.

Much of what we are learning about addiction and the effect of these substances on the brain comes from research applications of brain imaging techniques. In 1996 neuroscientists for the first time were able to use PET scans of the brains of cocaine addicts in the throes of craving to identify visually the neural bases of addiction. Imaging shows that when addicts feel a craving, there is a high level of activation in the mesolimbic dopamine system. In one study (Goleman 1996), PET scans were run on patients under treatment for cocaine addiction as they were exposed to cues associated with past craving episodes. The scans indicated activation of the dopamine system in the ventral tegmental area at the moment the addicts expressed intense craving. An Italian study (Tanda et al., 1997) found that the mesolimbic dopamine system was also active in nicotine addiction, while another study (Rodriguez de Fonseca et al., 1997) found that marijuana affected the same brain circuit. In addition to activity in the ventral tegmental area, these studies discovered similar activity in the outer layer of the nucleus accumbens and in the interconnected amygdala. The latter tie is supported by evidence that persons with lesions in a section of the amygdala are unable to link pleasure with its causes.

Moreover, this research is beginning to provide insights into how the drugs affect the brain. Studies of brain cells demonstrate that repeatedly exposing the brain to addictive drugs represents a chemical assault that alters the very structure of the neurons in the circuitry for pleasure. Over time these changes starve the affected cells of dopamine, thereby triggering a craving for the addictive drugs that will again activate release of high concentrations of that neurotransmitter (Goleman, 1996). During withdrawal, a different circuit in the same brain region releases a small protein, corticotropin-releasing factor (CRF). When a person suddenly stops taking the addictive substance, CRF levels rise and the person experiences withdrawal symptoms. Again, this process has been found to be identical for such addictive substances as nicotine, marijuana, alcohol, heroin, cocaine, and amphetamines.

In light of the technological advances of brain imaging and the current research on the biological bases of addiction to substances, it is likely that such research will be expanded to other addictions such as gambling, sex, eating, and so forth. Given what we know about the interaction of drugs and neurotransmitters in the pleasure circuits, it would not be surprising if similar effects were present with other pleasure-giving behaviors. The implications of this research for dealing with behaviors that are personally and socially destructive are, of course, substantial, as are the legal and policy ramifications.

There is a danger of extending the notion of addiction to any behavior which becomes patterned because it stimulates the pleasure centers. This notion has legal implications and raises questions again concerning free will and responsibility for one's own actions. The courts are going to be faced with novel defenses based on scientific evidence of genetic predisposition and neuronal susceptibilities. Evidence that all substances of abuse have similar impact on the brain despite differing mechanisms implies extension to nonsubstance factors that exhibit similar effects on the neural circuitry.

This evidence also has implications for drug policy that makes distinctions among potentially addictive substances (Pope and Yurgelun-Todd, 1996). Arguments for the legalization of marijuana become more difficult to accept given this evidence (Wickelgren, 1997b). Moreover, our society's treatment of nicotine and alcohol may have to be modified if consistency is sought. The evidence of the interchangeability of substances in producing similar effects on the activation of dopamine demonstrates that antidrug policies that focus on one drug at the exclusion of others are likely not to stem the addiction problem but rather to shift it to other substances when the supply of the first drug is cut. For the addict, it seems no drug is a safe drug, only a substitute. Neuroscience research on addiction, therefore, is likely to undercut some current policy initiatives and treatment regimes, but it offers the promise of more creative and effective solutions in the decades to come.

SEXUAL DIFFERENCES AND THE BRAIN

Until recently, human sexuality was the domain of psychology, but neuroscience findings are shifting emphasis to biochemical processes. This shift has naturally been criticized by those observers who hold that nurturing, culture, and social environment are the most powerful forces influencing sexual behavior as well as other behaviors and cognitive traits that typically differentiate the sexes. Current biological and neuroscience research, however, demonstrates that variation between the sexes and in sexual orientation are inextricably linked to differing hormonal influences on brain development. Although none of these findings eliminate environmental contributions to behavior, cumulatively they require a shift in balance from nurture to nature as a prime focus of inquiry. As a result, neuroscience is producing intensified conflict and shaking conventional foundations of our perceptions of sex differences and equality.

As with research in addiction, studies based on sophisticated brain imaging systems are providing dramatic evidence that male and female brains process information differently. Moreover, recent behavioral, neurological, and endocrinologic research indicates that the effects of sex hormones on brain organization occur so early in life that, from one's birth, the environment acts on differently wired brains in males and females (Kimura, 1992: 19). Furthermore, the biological factors that contribute to many sex-specific behaviors can be traced to both differing levels of sex hormones and sex-specific differences in the brain. The implications of these findings for public policy force us to conclude that "men and women have been living for the past thirty years with the absurd expectation that moral and political correctness demands gender sameness" (Nadeau, 1996: 14).

Brain Differences by Sex

There are three major areas of sexual differentiation: internal genitalia, external genitalia, and the brain. Although all three have genetic/

hormonal foundations, attention here focuses on sex differences in the brain. The evidence comes from extensive animal studies and more recent studies using PET- and MRI-based research on humans. Attention has centered on two major sites, the hypothalamus and the corpus callosum, which connects the two hemispheres, but other regions of the nervous and endocrine systems also exhibit differences by sex. It must be emphasized that the research findings reflect statistical averages and variations, and causality remains largely speculative, although evolutionary theories abound.

Male brains are on average larger than those of females, but this is mostly due to body-size differences. However, significant differences lie in the size and functioning of particular brain regions. The hypothalamus is a natural target for such research because it is the regulatory center of primal activities, including feeding, drinking, blood pressure, body temperature, growth, and emotional responses. This dime-sized region of the brain controls and modulates sexual behavior and is rich in androgen receptors. The hypothalamus is symmetrical, containing ten or so nuclei on each side. It is not only interconnected with the functions of the amygdala and hippocampus, but it also controls the secretory function of the pituitary gland. The sexually dimorphic nucleus of the hypothalamus is associated with sexual behavior, neural control of the endocrine glands, and sexual orientation. When a child is 2 to 4 years old, the release of testosterone promotes cell growth and prevents cell death in this nucleus, and as a result, it doubles in size in male children.

The region termed the medial preoptic area has been found to have a vital role in male-typical sexual behavior. This region has major hormonal inputs, especially testosterone, and incorporates several small nuclei as well as axonal tracts. When this region is destroyed in male animals, there is a cessation or reduction of copulatory behavior. Conversely, when this region is stimulated electrically, it has the opposite effect. Two of the nuclei in this region, INAH 2 and INAH 3, are on average larger in males than females.

In contrast, female-typical sexual behavior is modulated in a region slightly behind the medial preoptic area, in the ventromedial nucleus. Although it has also been linked to feeding behavior, this nucleus is associated with female copulatory behavior and is strongly influenced by sex steroids. In addition, sex-specific experiences have been isolated in the sirprachiasmatic nucleus in the hypothalamus, which is spherical in males but elongated in females. For both males and females, an intact hypothalamus is necessary for generation of sexuality, and puberty for both sexes is under its direct control through its complex circuitry with the neocortex and the amygdala.

Differences in metabolic activity in the brains of males and females have also been discovered through brain imaging studies. One study found seventeen regions of the brain where there were statistically significant differences in brain activity between male and female subjects at rest. Also, men on average have higher levels of activity in the temporal limbic system, a more primitive area of the brain associated with activity. In contrast, women have higher levels of activity than men in the middle and posterior cingulate gyrus, areas of more recent evolution and associated with symbolic action. Despite significant differences, there remains significant overlap in the sexes, however.

Although the hypothalamus is critical in explaining differences in sexual behavior between male and female subjects, recent studies have also found substantial differences by sex in other areas of the brain associated with nonsexual abilities, functions, and behaviors. The splenium, for instance, which is the back part of the corpus callosum, has been found to be larger in females than males. Because the corpus callosum connects the right and left hemispheres, if the actual number of fibers connecting the two hemispheres is larger, this could explain why female brain function is less symmetric than males. Communication between the hemispheres in females could thus be heightened because there are more routes connecting them. This could also help explain why damage to one hemisphere in a woman has a lesser effect than a comparable injury in a man (Kimura, 1992: 123).

The implications of these findings for the processing of information and for specific cognitive abilities are considerable. Research demonstrates that in part because of the greater interaction between the hemispheres of women's brains, the cognitive tasks of women tend to be localized in both hemispheres. In contrast, in males the two hemispheres act more independently, thus localizing cognitive tasks in only one hemisphere.

Language functions, for example, tend to be localized in different regions of the brain for men and women. One MRI study found that the male brain performs language tasks in the inferior frontal gyrus of the dominant hemisphere, whereas in females it takes place in both hemispheres (Shaywitz et al., 1995). Because females have a stronger concentration of left hemisphere linguistic function as well as more reliance on the right, they have a superior ability to learn complex grammatical constructions and foreign languages. The heightened interaction with the right hemisphere appears to enhance the range and complexity of linguistic representations in women. It might also help explain women's relative advantages in associational, expressive, and word fluency (Hyde and Linn, 1988).

The question of why male and female brains vary continues to be speculative, although the differences have been linked to exposure to sex hormones during the prenatal period. Kimura terms these effects as "organizational" because they appear to alter brain function permanently during a critical developmental period. Moreover, administration of the same hormones at later stages of development has no such effect, although cognitive patterns may remain sensitive to hormone fluctuations throughout life. "Taken altogether, the evidence suggests that men's and women's brains are organized along different lines very early in life. During development, sex hormones direct such differentiation" (Kimura, 1992: 125).

The prior question as to why such developmental differences exist is even more speculative, but one evolutionary perspective is widely discussed in the literature. Under this theory, sex differences in cognitive patterns arose because they proved advantageous. The assumption is that our brains, essentially unchanged over the last 100,000 years or so, reflect a division of labor in hunter-gatherer societies that put different selection pressures on males and females. Males were responsible for hunting, which required skills in long-distance navigation, the shaping and use of weapons, and spatial acuity. Women, in contrast, had responsibility for raising children, tending the home area, and preparing food and clothing. Their responsibilities required short-range navigation, fine-motor capabilities, and perceptual discrimination sensitive to small changes in the environment, skills that are consistent with findings of cognitive research. Moreover, men needed to be more aggressive for hunting and defense, whereas women required cooperative and consensual skills in the home and community.

Behavioral Differences by Sex

Whatever the ultimate cause of sex differences in the brain, they are reflected in varied cognitive capabilities and behavioral tendencies. In addition to the language skills differences discussed above, men on average perform better than women on certain spatial tasks, particularly those involving mental rotation. They also outperform women in mathematical reasoning tests, route navigation, and target-directed motor skills like throwing a baseball. Men and women in general construct three-dimensional space differently. Although women are stronger at verbal reasoning in mathematics, men are stronger in abstract mathematics. Other research has demonstrated that women, on average, are more skilled at hand-eye coordination, have better sensory awareness, have better night vision and wider peripheral vision, have longer attention spans, and are less likely to be either dyslexic or myopic.

Studies of infants have found that males are more interested in objects than people, are more skilled in throwing objects, and are better at following objects in space. By contrast, female infants are more interested in people's faces and voices and appear to be significantly more adept at assessing mood based on visual or voice cues. The games of girls place emphasis on cooperation and physical proximity, and they are anxious to integrate newcomers into the group play. Interaction is favored over specialized roles. Boys' games emphasize competition and action and favor clearly defined winners and losers. They are indifferent to newcomers and accept them only if they are useful. Girls have also been found to be better auditory listeners, whereas boys are better spatial-visual listeners.

One of the most studied differences between the sexes is aggression. Aggression has been found to be highly dependent on prenatal androgen exposure. Most research has focused on the role of the amygdala, especially the corticomedial and basolateral nuclei, which contribute to behavior that has a strong emotional loading such as aggression or fear-driven behavior. Destruction of the amygdala in animals leads to docile behavior. Moreover, studies of girls exposed to excess androgens during the prenatal period, who as a result have congenital adrenal hyperplasia, show that they grow up more aggressive than their unaffected sisters (Kimura, 1992: 122).

Male-typical behavior, therefore, demonstrates a strong bias toward action, heightened aggression, and command-oriented hierarchical structuring. Female-typical behavior, conversely, places more emphasis on consensus, cooperation, and interaction. While the male brain constructs reality in terms of vectors marking distance and space thus very segmentedly, the female brain tends to construct reality in terms of more extensive and interconnected cognitive and emotional contexts. As a result, females are more likely to feel a need to be included and attached, to share mutual feelings, and to receive confirmation of these feelings. Men, in contrast, remain more distant, unattached, and independent.

Neuroscience research, then, gives us new insights into why male and female interests, abilities, and worldviews are often at odds. It demonstrates that the influence of hormones on neural development is a powerful explanatory aide for patterns that have long been centers of controversy. Although considerable caution must be used in interpreting any of these data, the cumulative impact of this research on our understanding of human sexuality and of nonsexual differences between men and women is significant. Though these findings do not negate the importance of nurture to any individual's behavior or capabilities in these areas, they undoubtedly place learning in a much different context than has been the norm. For Nadeau:

> When men and women tend to solve problems differently, perceive different sets of relevant details, and display different orientations toward objects and movements in three-dimensional space, this is not merely learned behavior. These habits of mind are conditioned by sex-specific differences in the human brain. (1996: 12)

No wonder the findings of neuroscience and our knowledge of the brain's influence on behavior are not universally welcomed.

The Brain and Sexual Orientation

One of the most controversial findings of neuroscience centers on the role of the brain, in combination with the genes, on sexual orientation. In 1991 Allen and Gorski found differences in the size of the anterior commissure, the axonal connection between the left and right hemispheres. While the major finding was that on average it is larger in women than men, they also found that it is on average larger in gay men than either straight men or women, indicating that cerebral functions are less lateralized in gay men than straight men.

In a highly publicized extension of this study, LeVay (1991) scanned the cadaver brains of gay and straight men and of women assumed to be heterosexual. He focused his attention on the INAH 3 nuclei in the medial preoptic region of the hypothalamus, which is known to be sexually dimorphic, larger in males than females. As noted earlier, this region has major hormonal inputs and is characterized by high levels of androgen and estrogen receptors. LeVay found that on average the INAH 3 nuclei of gay men was the same size as those of the women and two or three times smaller than those of straight men. This finding suggests that gay and straight men may differ in central neuronal mechanisms that regulate sexual behavior. LeVay suggested two possibilities as to how this might come about. First, it could result from differences between gay and straight fetuses in levels of circulating androgens at the critical period for development of the INAH 3 nuclei. Or second, it could be that while levels of androgens are similar, the cellular mechanisms by which the neurons of INAH 3 respond to the hormones are different (LeVay, 1994).

Although LeVay concluded that both inborn and environmental factors influence the anatomical and chemical structure of the brain, there is much to recommend the theory that there are "intrinsic, genetically determined differences in the brain's hormone receptors or other molecular machinery that is interposed between circulating hormones and their actions on brain development" (1994: 127). While the factors that determine sexual orientation are not yet known, LeVay posits that it is "strongly influenced" by events occurring during the early developmental period, when the brain is differentiating sexually under the direction of gonadal steroids.

If there are indeed differences in the brains of gay and straight men, it is not unlikely that a gene or genes exert an influence on this process. It has long been known that homosexuality runs in families, but only recently has this been confirmed by twin studies. Bailey and Pillard (1991), for instance, found that if one identical twin is gay, the other is three times more likely to be gay than if the twins were fraternal. Having a gay maternal twin makes your likelihood of being gay about 50 to 65 percent, while the corresponding figure for a fraternal twin is about 25 to 30 percent. Other studies have found that having a gay brother increases one's chance of being gay to about 25 percent, as opposed to the total gay male proportion of the general population of 2 to 4 percent. In a comparable study of female twins, 48 percent of maternal twin sisters of lesbians were lesbians, while approximately 16 percent of fraternal twin sisters were lesbians (Bailey, Pillard, and Agyei, 1993).

There are three models that might explain these data of a genetic component of homosexuality: the direct, indirect, and permissive effect models. In the direct effect model, the genes influence the brain structures that mediate sexual orientation. In one approach of the direct effect, a gene directs a specific pattern of ribonucleic acid (RNA) synthesis, which in turn specifies the amino acid sequence of a particular protein, which in turn influences the behavior. Under the indirect effect model, genes code for personal factors such as temperament, which influences how the individual reacts with his or her environment. And finally, under the permissive model, genes influence neural substrates on which sexual orientation is shaped during the formative years. Although none of these models excludes an environmental component and the importance of many intervening pathways between genes and behavior, the direct model allows for less intervening influence (Schuklenk et al., 1997: 8) by assuming a more direct linkage between genes, hormones, and sexual orientation. In each model, however, the operative genes must be identified if the models are to move beyond speculation.

In 1993 Hamer and associates found hereditary linkages of gay patterns in the maternal line. This sex-linked pattern of inheritance suggested that a gene on the X chromosome might influence sexual orientation in men. They examined DNA from the X chromosomes of gay men and found a cluster of DNA markers at one end of the chromosome in a region called q28 that was statistically linked. Although a gene was not isolated, this evidence suggests that somewhere in the Xq28 region there is a gene or genes that predispose a man to be either gay or straight. There would be a 50 percent chance of getting two X chromosomes from the mother. This study has spurred even more interest in locating the gene, although its findings have been questioned by Marshall (1995) and others.

Although many gays, including LeVay and other researchers, welcome the evidence that homosexuality is genetic and neurologically based, some observers contend that in a homophobic prejudicial society it will have a strong negative effect on gays (Schuklenk et al., 1997: 12). As with other areas of genetic screening, there is a danger that presence of a gay gene or DNA marker will stigmatize the carrier or institutionalize the use of prenatal diagnosis and selective abortion of fetuses identified with the gene. Should such policies or practices be adopted, any potential gains that gays have in arguing that homosexuality is an immutable characteristic, a natural state like left-handedness, will be overshadowed by these practices.

Moreover, some have argued that the very research that attempts to find a gay gene or gay brain has a homophobic framework that will emphasize these traits not as natural polymorphisms but as dysfunctions or abnormal brain development (Schuklenk et al., 1997: 9). Even the motivation for seeking the origin of homosexuality is suspect, they argue. In the end, how society perceives and uses this information is political, and it will evolve in the broader social context to the extent that there remains latent or expressed homophobia in a society. Knowledge of the genetic/neural bases of sexual orientation will do nothing to stem discrimination against gays, even though it demonstrates that homosexuality is a natural state. Whatever findings emerge from science, the response to this knowledge depends on society and thus only indirectly through our collective neural connections.

Restak warns that we are now at the stage where "determination of individual responsibility is given over to neurologists and neuropsychologists and the concept of free will is relegated to the status of a museum piece dating from medieval scholasticism" (1994a: 152). Despite evidence of brain/behavior linkages, it should be noted that neural functioning remains a weak predictor of behavior. Even obvious cases of brain damage to the frontal lobes do not always lead to behavioral abnormalities or deficits. MRIs routinely identify abnormalities that have no discernible effect on the person's behavior. Moreover, most individuals with anti-social behavior exhibit normal brain functioning, as measured by current technologies.

For purposes of discussion, let us assume that a person does manifest severe deficits in neural functioning. Should this mitigate that person's responsibility for his or her behavior? As noted earlier, the legal system today makes exceptions based on concepts of diminished capacity. The question, of course, is where we draw the line. What is likely to confound the situation is the presence of sophisticated and precise means of mapping the brain and identifying deficits and our expanding knowledge of the role of neurotransmitters in explaining behavioral problems in biochemical terms.

A related issue is whether this new understanding of the brain's functioning undermines the concept of free will, or at least requires a redefinition of what it entails. Again, the issue is not a new one; there has always been disagreement as to whether an addicted person under the influence is legally and/or morally responsible for their actions. Similarly, sufferers of schizophrenia or other mental illnesses or neurodegenerating disorders are at least given a sympathetic hearing on the grounds that they are not responsible for their behavior. Our new knowledge simply expands the cases for consideration.

The fact that we are now aware that all expressions of what we view as the mind, including free will, are affected by the biochemical, electrical state of the brain should not, I believe, force us to abandon the notion of a free will, although it does require a refinement of it. For all the growing evidence on the crucial role of the brain (and genes) for behavior, it is rarely determinative. Alcoholics under twelve-step programs can and often do refrain from drinking. Although they do not attribute this to will, certainly it plays a critical role in some form. Even some pedophiles have been known to control their compulsions.

Despite the knowledge of neuroscience, humans do retain the capacity to make conscious decisions—this is what continues to separate us from other mammals. The notion of free will is still functional, although the traditional notion based on a tabula rasa has long been outdated. Although the demise first of the soul and then of the mind are troubling to many, they do not signal the end of ultimate individual responsibility for actions.

That we are not fully rational, entirely conscious creatures whose actions are determined solely by logic and reason should come as no surprise. Mr. Spock is not of the world of humans, even though Descartes might have wanted him to be. Humans are constrained by brains that have evolved from primitive times during which emotions of fear and aggression were crucial to survival. Although free will in an absolute sense is, and most likely always was, a philosophical artifact with little grounding in reality, it remains relevant though qualified, at least at the margins. With few exceptions, individuals ultimately bear responsibility for their own actions.

Ironically, a more major threat to free will and individual responsibility comes from the opposite perspective. Recently it has been argued by persons with an environmentalist bent that individuals should not be held responsible for actions beyond their control. The control implied here, however, refers not to genes or neurons but to the social context. For instance, it is suggested that to place responsibility for ill health on those persons who behave in ways that cause their illness is to "blame the victim" (Fitzgerald, 1994). As stated by Etzioni, "health as individual responsibility . . . tends to overlook or misconstrue the nature of the social constraints on the individual will" (1978: 65).

Individuals, in this perspective, are incapable of making rational decisions when it comes to their health. Peer pressure, media advertising, social and economic constraints, and cultural forces in effect create an incentive structure that shapes and limits a person's own choice. Thus, to hold the person responsible, to invoke the possibility of free will, is unfair according to these critics of individual responsibility. This argument comes from both ends of the ideological spectrum. Although the most vocal proponents of the socially determined model of behavior come from the left, similar views are espoused by conservatives as well. Fundamentalist religious leaders rail against a society that perpetuates promiscuity, pornography, and so forth through the mass media and assume that impressionable individuals are powerless to exercise their own judgment because of these things.

The concepts of free will and individual responsibility, then, face pressures not only from emerging knowledge in genetics and neuroscience, but also from changes in social values that would isolate individuals from personal responsibility for their decisions. As already noted, although free will is a basic assumption of our legal and political system, disparate forces are slowly undermining our faith in the independence of individuals to exercise free will. Although genetic and neurological findings are likely to contribute to this process, they, themselves, must be placed in the broader perspective of changes in social values that appear to have a momentum of their own. The biggest danger is that genetic and neurological arguments will be used to reinforce the view that free will is no longer relevant.

CONCLUSIONS

The implications of neuroscience for the study of political behavior and politics in general are extensive. On the practical level, knowledge of neuroscience has ramifications for public policy, as well as for how we structure elections, conduct politics, and organize bureaucracies. Controversy over the nature/nurture debate, social stratification, and antisocial behavior is bound to escalate in light of findings in genetics and brain research. Our new knowledge of the biological bases of behavior also has significant implications for the justice system.

Normative political theory will have to confront the challenges of life science research at three levels of analysis: the individual, intrasocietal, and extrasocietal. Cognitive neuroscience challenges the conventional view of mind/body dualism, rational choice, tabula rasa, and the status of human consciousness, all of which have ramifications for our conceptions of human nature and thus of the individual in the state. It has substantial implications as well for the study of human aggression at all three levels of analysis. Long-held assumptions of democracy must be reevaluated in light of these challenges. Likewise, the concept of democracy itself and assumptions underlying equality require redefinition.

It is therefore not surprising that biological models of political behavior are so controversial. Challenges to liberal theories are also inherent in biological findings that demonstrate that the division of personality into conflicting elements of reason and emotion are inappropriate. Rational voter models and rational choice theory, for example, assume that reason is the calculating mechanism and that passion is personal and dangerous in political decision making. Full theories of politics based on brain function must synthesize both reason and emotion and deal with an integrated human personality, contrary to existing theories at the base of political behavior and political decision making.

Finally, empirical political theory and methodology must be reexamined to account for this enlightened knowledge of human behavior. Subsequently, methodologies must be adapted to allow for a life sciences model, as opposed to the conventional Newtonian physics model. Additionally, health and illness must be incorporated into survey research and appropriate measures of health status must be operationalized. Biological and psychological variables should supplement political opinion and attitude scales if we are to understand more fully political behavior. This will entail the development of new experimental designs to supplement conventional political research.

6

Brain Intervention Techniques

As noted in Chapter 1, we are currently witnessing a rapid expansion in the capacity to intervene in the brain and nervous system. Although many of the techniques are recent, the issues surrounding their use are not. One continuing issue centers on the use of certain techniques for behavioral disorders or antisocial behavior. In the absence of clear evidence of organic brain damage, intervention in the brain—whether electrical, physical, or medical—raises concerns about personal autonomy, safety, and efficacy. These interventions are also questioned as to whether treatment of symptoms instead of cause does not simply replace one problem with others. The development of innovative methods of treatment of mental, psychological, and behavioral disorders will not resolve these continuing issues surrounding attempts to intervene in the brain. This chapter reviews the current state of electrical, physical, and chemical interventions and the specific policy issues they raise.

PHYSICAL INTERVENTION IN THE BRAIN

Direct electrical or physical interventions in the brain clearly illustrate the issues introduced in Chapter 1, and they will continue to elicit strong opposition even where deemed safe and effective. Because of the unique nature of the brain and its relationships to autonomy and personality, interventions, whether electrical or surgical, are viewed by many people as assaults on personhood. Direct brain intervention also raises questions concerning the line between therapy and experimentation because of the uniqueness of each brain. Furthermore, questions of consent are especially troublesome when the intervention is potentially irreversible or when the patient is deemed ill enough to undergo such a procedure. And finally,

105

these techniques all raise the potential for abuse and the specter of behavioral control that critics generally exploit.

Electroconvulsive Treatment (ECT)

Anyone who has seen *One Flew over the Cuckoo's Nest* would be hard-pressed to forget the electroconvulsive treatment (ECT) scenes, in which the patient is secured and a series of 70 to 130 volts of alternating current are administered to the nervous system for one-tenth to one-half of a second, resulting in a series of spasmodic muscular contractions. Although electroconvulsive treatment is viewed by many persons as a very crude form of therapy, in 1989 the American Psychiatric Association (APA) called the treatment "safe and very effective for certain severe mental illnesses," including major depression, bipolar disorders, and schizophrenia (Staver, 1989). ECT should be used only after standard treatment alternatives have been considered and psychiatrists using it have a "serious obligation" to obtain the informed consent of the patient. Although in the past the patient was awake for treatment, now the patient is given a short-acting anesthetic and an injection of a strong muscle relaxant before the current is applied. After several minutes the patient awakens, remembering nothing about the treatment and experiencing no pain, although the patient does experience a short period of disorientation. The usual ECT regimen entails about four to twelve treatments given every two days over a period of several weeks. An estimated 100,000 persons undergo ECT each year in the United States (Winock, 1997: 88).

Although ECT generally relieves symptoms of depression rapidly, it is not without complications. The most common side effect is memory loss for an indeterminate period of time following the procedure. Memory dysfunction can affect both the capacity to recall material learned before the procedure (retrograde) and the capacity to learn new material (antero-grade). In most cases, anterograde memory gradually improves after treatment. Likewise, retrograde memory capacity usually reaches full recovery within seven months after treatment, although in many cases subtle memory loss will continue beyond that time. In very rare cases, such persisting loss is comprehensive and debilitating (Bootzin et al., 1993: 564). Also, because many patients are fearful of ECT, it is and has been abused as a means of behavioral control when used as a threat to coerce cooperation. Its shotgun-type effect and its violent image, combined with a history of abuse in some institutions, have made ECT a most controversial form of intervention in the nervous system.

According to Abrams, the ill-conceived and large-scale misuse of ECT from the 1940s to the 1960s damaged its image (1988). The fears and prejudices of ECT by both the public and physicians were heightened by its use without anesthesia and by unqualified personnel (Durham, 1989). In recent years, at least thirty-seven states have restricted the conditions under which ECT can be provided. For Petersen, regulations controlling ECT are so stringent that many of the most direly ill patients now have little or no chance of receiving this treatment, despite its approval by the APA and overwhelming evidence of its safety and efficacy (1994: 29–30). Not only does suppressive regulation of ECT deprive patients of treatment, but it raises a number of policy questions regarding the health interests of the community, self-determination, competency, and privacy. For Petersen, although the phenomenon of involuntary treatment represents a tiny portion of ECT use, it has led to sensationalization and polarization and has denied the preponderance of patients a free choice under the guise of protective legislation (1994: 30). Comparatively, critics of ECT argue that it is a most obvious form of psychiatric assault on mental patients, and they equate it with strong negative images like those presented by Hollywood.

Electronic Stimulation of the Brain (ESB)

Electronic stimulation of the brain (ESB) is likely to replace ECT as an electronic means to ameliorating behavioral disorders. ESB does not induce a permanent beneficial change but instead an emotional tranquility. It is also effective in relieving psychological pain caused by severe anxiety and depression by evoking a feeling of euphoria. ESB consists of thin insulated wires implanted in the brain that allow electronic messages to be sent and recordings to be made from areas deep inside the brain. (For therapeutic reasons, the most common placement is in the septal region.) ESB has also been used successfully in the treatment of epilepsy, intractable pain, violent aggressiveness, and chronic insomnia. Evidence to date suggests that ESB causes no permanent anatomical damage to the brain and that its effects are reversible, but its continuing experimental nature warrants caution in its therapeutic applications.

In August 1997, the Food and Drug Administration approved the use of the deep-brain stimulator for the treatment of tremors caused by Parkinson's disease and essential tremor disorder, which together affect approximately 3.5 million Americans. The Activa brain implant involves drilling through the skull to implant an electrode in the thalamus, deep inside the brain. A wire runs down under the scalp to the collarbone, where a tiny pulse generator is implanted. This device sends waves to the electrode, which in turn emits constant, tiny electrical shocks that block the tremors. The pulse generator can be adjusted by radio wave or turned off by running a magnet over the generator. In an initial study of 196 patients with severe Parkinson's disease and essential tremor, 67 percent of the former and 58 percent of the latter experienced significant reduction in tremors, though about 5 percent of the patients suffered bleeding in the brain or other complications (Neergaard, 1997). As we come to better understand the linkage of specific nuclei and areas of the brain to debilitating disorders, electrical stimulation systems will become more numerous.

When used to treat abnormal aggressiveness, periodic ESB is required. Once the electrodes are implanted, brain activity can be monitored and warning signs of an impending crisis identified. This diagnostic capacity might be the only means of revealing abnormal patterns of electrical activity deep inside the brain. The development of miniaturized electronic devices—stimoceivers—permit instantaneous radio transmission to and from the brains of subjects. Computers are used to permit remote monitoring of brain activities—they also allow complete control over the brain of the subject, raising severe ethical questions. For instance, some observers have suggested the use of such devices on parolees to monitor their activities, movements, and brain waves by computers in centralized locations (much like an air traffic control system). The opposition to such a proposal has been intense. ESB, however, clearly demonstrates that even a technique that has credible uses when medically indicated has tremendous abuse potential. By allowing complete control over the general mood of the subject, ESB raises basic concerns of personal autonomy and self-determination.

Psychosurgery

Because it results in permanent destruction of particular regions of the brain, the most controversial type of physical brain intervention is psychosurgery. Neurosurgery is a credible and accepted therapeutic procedure. When used to repair physical damage caused by head injuries or tumors, brain surgery raises few objections. When the same techniques are used instead to correct mental and/or behavioral disorders (psychosurgery), they become highly controversial. These applications rest on the assumption that behavioral disorders have an organic base and that to treat a disorder, the organic pathology must be corrected by physical intervention—not psychotherapy. This assumption is highly debatable, according to Chorover (1981) and other critics of psychosurgery. Furthermore, the short history of psychosurgery is not one to inspire confidence.

According to Valenstein (1986: 3), between 1948 and 1952 "tens of thousands of mutilating brain operations were performed on mentally ill men and women," including many from the United States who voluntarily underwent lobotomy. Although the practice of lobotomy was curtailed by 1960, primarily because of the availability of psychoactive drugs as an inexpensive alternative, in its wake it left many seriously brain-damaged persons. Bootzin and associates assert that lobotomy is now resorted to only rarely (1993: 565).

Frontal lobotomies entail cutting the frontal lobes off from the rest of the brain, thereby making them inoperable. Although several techniques were used to lobotomize approximately 70,000 persons in the late 1940s and early 1950s, Walter Freeman achieved critical acclaim for his use of transorbital lobotomy. This technique is casually summarized by Freeman in its gruesome details:

> I have also been trying out a sort of half-way stage between electroshock and prefrontal lobotomy on some of the patients. This consists of knocking them out with a shock and while they are under the "anesthetic" thrusting an ice pick up between the eyeball and eyelid through the roof of the orbit actually into the frontal lobe of the brain and making the lateral cut by swinging the thing from side to side. I have done two patients on both sides and another on one side without running into any complications, except a very black eye in one case. There may be trouble later on but it seemed fairly easy, although definitely a disagreeable thing to watch. (cited in Valenstein, 1986: 203)

Interestingly, the pejorative term "ice pick surgery" is accurate. Although Freeman eventually developed the "transorbital leucotome" and then the "orbitoclast" as the instruments designed to withstand the tremendous pressures necessary to crush through the orbit before severing the brain, the first transorbital lobotomy was performed in his office using an ice pick with the name Uline Ice Co. on its handle.

Although a large proportion of lobotomies were performed in state hospitals on indigent patients, a substantial number were performed on well-to-do women in private hospitals and physicians' offices. The transorbital procedure left only small scars on the eyelids and could be performed in a matter of minutes. To some extent, frontal lobotomies took on a fadlike image. The fact that such a gruesome and intrusive technique won not only uncritical acceptance by the medical profession as a whole (despite a few vocal detractors) but also widespread public approval and mass media support as a miracle cure reiterates the faith society has in a technological fix. As discussed later, this dependence on technology was easily transferred to the less dramatic psychoactive drug therapy common after 1960.

The demise of lobotomies did not end psychosurgery. Sophisticated stereotaxic instruments facilitate more precise placement of miniature electrodes on specific brain targets using geometric coordinates under imaging and thus allowing for the destruction of relatively small areas of brain tissue. Electrolytic lesioning or selective cutting of nerve fibers is conducted after the region to be lesioned is localized by establishing coordinates using anatomical landmarks and X-rays. In one procedure, called stereotaxic subcaudate tractotomy, a small localized area of the brain is destroyed by radioactive particles inserted through small ceramic rods. The particular site varies with the nature of the disturbance. "For depressed patients, it is in the frontal lobe; for aggressive patients, it is the amygdala, a structure in the lower part of the brain" (Bootzin et al., 1993: 565). There have also been some applications using radiation, cryoprobes (freezing), and focused ultrasonic beams to destroy the tissue.

As with lobotomies, however, there is no way of predicting the consequences of these procedures in either the short or long run. According to critics, psychosurgery is a highly experimental procedure that "produces a marked deterioration in behavior, serious impairments of judgment, and other disastrous social adjustment effects" (Chorover, 1981: 291). Not surprisingly, Chorover has called for a temporary moratorium on psychosurgery and for more basic research on brain mechanisms and behavior. In contrast, Valenstein (1986: 285) contends that after a lag in psychosurgery since the mid-1970s, new knowledge about the brain, combined with the development of better surgical techniques, seems to have justified reconsideration of psychosurgery for patients who do not respond to other treatment.

Although the National Commission for the Protection of Human Subjects (1976) criticized the use of psychosurgery for social or institutional control, it concluded that it might help some patients as currently practiced. Though it recommended prior screening of each proposed surgery by an independent institutional review board (IRB), its recommendations were never translated into federal legislation or regulations. Instead, action restricting psychosurgery has originated in state legislatures. Oregon, for instance, passed a statute in 1982 that strengthened its earlier restrictions (1973) and prohibited psychosurgery in the state. California also has instituted regulations that have all but ended the practice. Although most states have not taken action, primarily because of the threat of liability, less than 200 psychosurgeries are performed annually in the United States.

Ironically, concurrent with the imposition of legal restraints, extensions of studies originally reported to the National Commission have been favorable to psychosurgery for certain classes of patients. Two independent research teams (Teuber, et al., 1977; Mirsky and Orzack, 1977) reported that the quality of life of between 70 and 80 percent of patients improved significantly after undergoing psychosurgery. Furthermore, they found no evidence of physical, emotional, or intellectual impairment caused by the surgery. Careful patient selection and use of surgery as a last resort only for those patients who fail to respond to drug therapy seem central to any future uses of psychosurgery.

Even under tight limits, however, psychosurgery engenders significant debate. In *Violence and the Brain* (1970), Mark and Ervin raised intense controversy by charging that much of the violence in the United States is caused by brain pathology. According to these authors, patients with "discontrol syndrome" were prone to sudden violent outbursts triggered by abnormal electrical discharges in the temporal lobe. They recommended an ambitious effort to locate the triggers of these outbursts and remove them. This call for preventive brain surgery as a means of dealing with a social problem directly raised the question of social control of behavior that is judged troublesome or abnormal.

In opposition, Thomas Szasz (1974) contends that diagnostic labels in psychiatry are myths used to stigmatize social deviants. With its use of involuntary confinement, mind-altering drugs, and psychosurgery, psychiatry for Szasz has become a political force, concealing social conflict by calling it illness or justifying coercion as treatment. Chorover (1981: 291) agrees that although psychosurgery has unique characteristics, in terms of social policy it is merely one of a large number of "psychotechnological" means of dealing with individuals or groups vaguely defined as aggressive, disruptive, dangerous, uncooperative, and so forth.

Insofar as the causes of social conflict actually lie in the domain of social affairs, psychotechnological treatment of deviants should be regarded as a perversion of medicine and a distinct threat to individual liberty (Chorover, 1981: 291). Chorover agrees that we must assess the impact and social consequences of psychotechnology within the broad context of politics and public policy; otherwise, we surrender both our constitutional freedom and our human dignity.

Psychosurgery, then, continues to generate policy issues on a range of levels. The first issue is the dilemma of consent. As noted earlier, the consent must come from the damaged organ itself. But if the damage is severe enough to warrant surgery, who can give consent for an irreversible procedure? Second is the question of therapy versus experimentation. Because every person's brain is unique, psychosurgery will always have a high degree of uncertainty and thus risk. Third is the question of assuming that deviant behavior has an organic base. If it does not, can we justify the use of organic procedures such as surgery to resolve nonorganic problems? Are we not simply treating (i.e., controlling) the symptoms instead of dealing with the cause?

Psychosurgery also vividly illustrates problems of social control. We are developing an impressive array of techniques to control or modify behavior. Each technique, however, offers tremendous opportunity for abuse and poses serious threats to individual liberty. The stigmatization associated with being labeled abnormal, the constraints on freedom of choice, and the erosion of the dignity of the individual are seriously challenged by the presence of these innovations. In contrast, there is evidence that ECT, ESB, and psychosurgery may be beneficial to many individuals and that for some persons they represent the only hope of leading a near-normal existence. To deny them the benefits of these technologies because we do not trust our ability as a society to set reasonable limits on their use seems unfair.

DRUGS AND THE BRAIN

Though less dramatic than physical interventions in the brain, the development of powerful, specific, psychotropic drugs cumulatively represents a much larger impact on the population. Psychopharmacology, the study of the treatment of psychological disorders with drugs, has exhibited remarkable advances in the last two decades and represents the mainstay of neuroscience research. New drugs are introduced to the market each year, often with great enthusiasm and fanfare, for treatment of the entire range of mental disorders, from schizophrenia and panic and obsessive-compulsive disorders to personality disorders and stress-related physical disorders. A new generation of drugs for enhancement of brain capabilities and treatment of neurodegenerating diseases promises to significantly expand medical intervention into the nervous system. Although some of these drugs fail to have the effects hoped for or are accompanied by adverse side effects that limit their use, overall the research in psychopharmacology has undoubtedly contributed to our understanding of the chemistry of the nervous system and has extended substantially the range of treatment possibilities.

Despite this record of advancement, drug treatment raises many problems. Many of the same issues triggered by physical brain intervention are mirrored in chemical intervention. Other than the irreversible dimension of psychosurgery, the use of drugs to produce the desired mood and mental functioning and thus influence behavior raises identical policy issues and makes similar assumptions on the organic base of behavior problems. Drug treatment for behavioral and psychological problems also brings to the forefront important issues as to whether it is wise to treat symptoms without addressing the underlying cause and under what circumstances individuals with severe mental disorders can be coerced into taking drugs that stabilize them but do not treat the cause.

Chemical control has considerably more policy importance because it enjoys such widespread and socially acceptable usage. The ease of its administration and the potential for surreptitious applications make this form of control more threatening than physical methods.

Psychotropic Drugs

There are three major groups of psychotropic drugs used as therapeutic agents. The first and most powerful are the antipsychotic drugs or major tranquilizers (see Table 6.1), which are used to relieve symptoms of psychosis including withdrawal, delusions, hallucinations, and confusion. With the FDA approval and marketing of chlorpromazine (Thorazine) in 1954, the first of a group of powerful phenothiazines was introduced to be used in the treatment of major mental illnesses such as schizophrenia and paranoia. These drugs demonstrate sedative, hypnotic, and mood-elevating effects. Although they do nothing to cure the disease, but rather suppress the symptoms, they are effective in maintaining equilibrium for many patients, allowing them in many cases to be released from

TABLE 6.1
Antipsychotic Medications

Chemical family	Generic name	Brand name
Phenothiazines	chlorpromazine	Thorazine and others
	thioridazine	Mellaril, Millazine
	mesoridazine	Serentil
	trifluoperazine	Stelazine, Suprazine
	perphenazine	Trilafon
	fluphenazine	Prolixin, Permitil
	triflupromazine	Vesprin
	prochlorperazine	Compazine
	acetophenazine	Tindal
Butyrophenones	haloperidol	Haldol
	pimozide	Orap
	droperidol	Inapsine
Thioxanthenes	thiothixene	Navane
	chlorprothixene	Taractan
Dibenzoxazepine	loxapine	Loxitane
Dihydroindolone	molindone	Moban

Source: OTA, 1992: 53.

institutions. Maintenance therapy requires prolonged use generally, how-ever, at reduced dose levels. Discontinuance of the drug will result in the return of symptoms. The major tranquilizers have also been used in treating violence. Yet administration is complex because effects and dosage levels vary across individuals and even within an individual patient.

Not surprisingly, these powerful antipsychotic drugs have serious potential side effects, including fatigue, apathy, blurred vision, constipation, muscle rigidity, and tremors. The most serious medical side effect of these drugs is tardive dyskinesia, a muscle disorder in which patients grimace and smack their lips incessantly. Unlike other side effects, this condition does not abate when the use of the drug is withdrawn, and it has proven resistant to treatment by additional drugs, unlike many of the other side effects.

Two other issues surround the use of antipsychotic drugs. Because of their tranquilizing effect, they have been used in some institutions in high doses for the management of patients who are disruptive. This abuse for behavioral control purposes should not detract from the benefits of antipsychotic medication, but it warrants close monitoring of its use under such conditions. The other issue, which is discussed later in more detail, is that patients released from institutional care into the community under the calming influence of a drug often stop taking that drug and must be readmitted. In other cases, these patients create problems in the community, and their disruptive behavior results in calls for coerced treatment or reinstitutionalization.

The second major group of psychotropic drugs are the antidepressant drugs, which are used to elevate mood in depressed patients. This class of drugs includes amphetamines, monoamine oxidase inhibitors (MAOIs), tricyclic derivatives of imipramine, and serotonin-reuptake inhibitor antidepressants such as Prozac (Table 6.2). Although amphetamines have little clinical value in the treatment of depression because of the very short duration of their effect, they are widely known and used. They also have considerable potential for abuse and dependency.

Until the introduction of Prozac to the market in 1987, the most commonly prescribed antidepressants were tricyclic antidepressants (TCAs), which produce an increase in the neurotransmitters norepinephrine and serotonin in the nervous system by blocking their reabsorption by the nerve cells. MAOIs generally have been used for patients who fail to respond to tricyclic antidepressants, but they can have serious side effects, including cardiovascular and liver damage. In combination with certain foods and medications, they can produce fatal bouts of hypertension (OTA, 1992: 58). Both inhibitors and tricyclics have side effects similar to those of antipsychotic drugs, including drowsiness, blurred vision,

TABLE 6.2
Medications for Treatment of Depression

Class of medication	Generic name	Brand name
Tricyclic antidepressants	amitriptyline	Elavil, Endep
	nortriptyline	Aventyl, Pamelor
	protriptyline	Vivactil
	desipramine	Norpramin, Pertrofan
	doxepin	Adapin, Sinequan
	imipramine	Tofranil, Imavate
Inhibitors	tranylcypromine	Parnate
	phenelzine	Nardil
Newer antidepressants	fluoxetine	Prozac
	sertraline	Zoloft

Source: OTA, 1992: 58.

nervousness, and constipation, but normally these effects are milder with antidepressants due to the lower dosages used. Another problem with these drugs is the lag time between their administration and their clinical effect. Although they increase neurotransmitter levels almost immediately, their therapeutic effects often do not appear for two or more weeks, a long time for a severely depressed patient to wait.

In the late 1980s the new generation of antidepressants termed selective serotonin-reuptake inhibitors (SSRIs) was marketed with the promise of offering safe and effective treatment of a wide variety of disorders. Over the past decade, SSRIs have assumed the role of first-line antidepressant agents. According to Kamil, the basis for this rapid acceptance includes "an increasing interest in the serotonergic system in the neurobiology of depression. . . , the extremely good safety profile of SSRIs in overdose, and the lower incidence of side effects compared with TCAs and classic MAOIs" (1996: 163). In light of an increasing public awareness of depression and an aggressive marketing strategy by Eli Lilly and Company, by 1990 Prozac was hailed as the antidepressant of choice. Because it was designed specifically to block serotonin reuptake, it has an antidepressant action equivalent to the older drugs but without many of their side effects (Rickels and Schweizer, 1990).

Shortly after its well-publicized introduction, Prozac became the most widely prescribed antidepressant in the United States. Because it keys in only on serotonin, Prozac is effective for many depressed as well as obsessive-compulsive patients who are unresponsive to MAOIs or tricyclics. Normally its side effects are limited to headaches, upset stomach, and anxiety. However, because it has been linked to extreme "behavioral toxicity," including manic agitation, violence, and suicide in some users, Prozac has become controversial. In several well-publicized cases, it has been used as an attempted defense in criminal trials. In July 1991, the FDA rejected a petition that sought a ban of Prozac, and later that year an FDA advisory panel found no evidence to link Prozac use with suicide or violent behavior. As with all psychotropic drugs, it appears that Prozac, although of great benefit to many patients, carries with it potentially significant risks for some users.

Despite this, Prozac and other SSRIs have been used to treat a growing variety of ailments, from depression and obsessive-compulsive behavior to severe premenstrual syndrome and even shyness. In 1997 the use of Prozac as a diet drug emerged. NutriSystem Weight Loss Clinics substituted a combination of Prozac and phentermine for the banned drug combination of fenfluramine and phentermine (fen/phen).

Another drug that has generated considerable controversy is Ritalin, a stimulant that has been found to have the opposite effect on hyperkinetic children. Because of its ability to calm hyperactive children, it has been documented that Ritalin has become overused in some school districts to pacify disruptive children, leading to frequent lawsuits (Moss, 1988). In the United States, Ritalin has also become the drug of choice for abuse among high school students. The use of Ritalin in schools for social control has come under considerable attack in recent decades, in part because the effect of prolonged exposure to it is unknown. Ritalin has, in fact, been linked to complex changes in the central nervous system. Questions of its effects on personality development and innovative thinking capacity, and of any psychological or physiological dependence on it, are unanswered. According to Safer and Krager (1992), since 1987 use of Ritalin by schools has declined due to a negative media blitz, lawsuits threatened or initiated, and the apprehension of parents and involved professionals.

The third grouping of psychoactive drugs are the antianxiety drugs or minor tranquilizers (see Table 6.3). These include barbiturates, which are the least effective and have a high tendency to produce dependence, habituation, and addiction. Despite these factors and the drugs' low margin of safety, barbiturates such as phenobarbital are effective in treating epilepsy. Another group of minor tranquilizers includes diazepoxides such as Librium and Valium. These drugs are used effectively to control muscle spasms, hysteria in acute grief situations, and compulsion. They are less dangerous and less addictive than barbiturates. Minor tranquilizers are not effective in the treatment of psychoses but are of special value

TABLE 6.3
Antianxiety Medications

Chemical family	Generic name	Brand name
Benzodiazepines	chlordiazepoxide	Librium
	diazepam	Valium
	chlorazepate	Tranxene
	oxazepam	Serax
	lorazepam	Ativan
	alprazolam	Xanax
Propanediol carbamates	meprobamate	Miltown
Azaspirodecanediones	buspirone	BuSpar

Source: Bootzin et al., 1993: 557.

in the treatment of tension and anxiety associated with situational states and stress.

The long-term use of antianxiety drugs is discouraged, however, because they can produce dependence, and they have a tendency to be abused. Moreover, antianxiety drugs have side effects including fatigue, drowsiness, and impaired motor coordination. They have been linked to automobile and industrial accidents and to falls in the elderly, who as a group represent heavy users. Antianxiety drugs can be dangerous if combined with nervous system depressants such as alcohol. Benzodiazepines, like barbiturates, have a synergistic effect when combined with alcohol: each multiplies the other's power, placing the person at risk for toxic overdose (Bootzin et al., 1993: 556). Barbiturate/alcohol combinations have lead to brain death in well-publicized cases such as that of Karen Quinlan.

Antianxiety drug use also risks the effect of rebound once treatment is terminated. Rebound means that the symptoms return with redoubled force, thus the person is likely to resume taking the drug at a higher dose in order to suppress the intensified symptoms. In effect, the patient becomes dependent on the drug at higher and higher doses. Antianxiety drugs are also criticized because they suppress the anxiety that is a signal for a more basic problem. The drug allows the taker to avoid facing the problem, thereby risking the problem to become even more serious, leading to the "need" for more drugs to relieve the anxiety, and so on.

In addition to the psychotropic drugs, the use of hormonal treatment to control behavior has proceeded rapidly since the 1970s. Studies have repeatedly found that aggression is highly dependent on testosterone levels. Elevated levels of this male hormone have been implicated in crimes of sex-related violence. The administration of female hormones to control aggressive males has been successful in treating abnormal sexual preoccupation. Depo-Provera has also been used to inhibit the male sex drive. In large doses Depo-Provera serves, in effect, as a chemical castrator by shrinking the testes (Green, 1986).

Drugs are used to alter behavior because they are effective and convenient, not because of a compelling scientific consensus as to how they help patients. In other words, there continues to be a use of technology for nonmedical reasons in order to accomplish other policy objectives. In a rights-oriented society, this practice raises serious constitutional questions. However, in our drug-oriented society, psychoactive drugs are routinely accepted to help us cope with day-to-day problems. To a large extent, drugs have become a quick fix for the anxiety, depression, and social stresses of modern-day existence.

Enhancement Drugs

Recently this fascination with pharmaceutical solutions has focused on the development of "smart" drugs, or nootropics, designed to increase brain power and to improve memory, concentration, and the ability to learn (Dean and Morgenthaler, 1992). In general, nootropics are compounds that act in the central nervous system, enhancing the cognitive performance by improving memory, perception, attention, judgment, and orientation. Although some nootropics are now available by prescription, in many cities in the United States such drugs can be purchased in smart drug bars (Whitehouse et al., 1997: 15).

Despite considerable variation in chemical composition and in the mechanisms by which they act, a common property of nootropic drugs is their effect on higher integrative functions of the brain. The more we understand about the actions and functions of specific neurotransmitters, the more likely we are going to be able to design drugs with nootropic characteristics. Windisch (1996: 242) cautions that while spectacular improvement has been documented in specific individuals, currently the average efficacy of nootropic drugs is usually low—10 to 20 percent as compared to a placebo—and side effects are commonplace. He argues that while no single substance is likely to cure a demented patient by cognitive enhancement, given the dramatic impact that Alzheimer's disease has on patients and society, even moderate reductions in symptoms or short delays in disease progression make research into nootropics important (Windisch, 1996: 251).

Although the original rationale for research on these substances centered on the goal of treating patients with premature dementias such as Alzheimer's disease, increasingly they are being touted as means of promoting mental agility in healthy persons who want to boost their intelligence. A "smart drug" industry is flourishing, despite the lack of scientific evidence that these substances actually work as performance enhancers in normal persons (Concar and Coughlin, 1993). The appeal of a technological shortcut to learning is apparently widespread (Rose, 1993).

This appeal is not limited to nootropics. It is likely that many drugs introduced to treat diseases have the capability to enhance certain nervous system functions. As we come to understand more about neurotransmitters, this likelihood will increase. One interesting application of this involves the use of beta-blocking drugs such as propranolol. Propranolol was first used for the treatment of cardiac arrhythmias and hypertension and to prevent sudden death after myocardial infarction. Its use as a treatment for migraine headaches and stress-induced anxiety followed shortly. Beta-blocking drugs work because they compete with adrenaline-like chemicals produced by the sympathetic nerves that attach to beta-adrenergic receptor sites when the body is under stress. The physiological effects of the adrenalinelike chemicals include rapid heart beat, muscle tremor, dry mouth, perspiration, and nausea. By occupying the receptor sites, propranolol blocks these physiological responses, thus reducing the symptoms of anxiety.

Although prescription of beta-blockers to relieve clinically diag-nosed anxiety is largely accepted, questions have arisen over its use to enhance performance (Salmon, 1990). One study found that 27 percent of the members of the International Society of Symphony and Opera Musicians report using beta-blocking drugs, and that 70 percent of that amount said they had used them without a prescription. Slomka (1992) contends that nontherapeutic use of such prescription drugs to enhance performance raises questions of unfair advantage and suggests that when one person is perceived as having such an advantage, others will be compelled to use it. "To the extent that such drugs actually confer a competitive advantage, their use by some people will result in pressure on nonusers to become users, or else to accept what amounts to a handicap in the social competition" (Whitehouse et al., 1997: 20).

Although the use of enhancing drugs might make for better perfor-mances, such a practice again demonstrates a willingness to use drugs to override normal feelings, emotions, and physiological processes. Though users and their supporters might interpret this repression as enhancing their autonomy by enabling the performer to exercise control, the use of beta-blocking drugs for normal levels of anxiety might alternatively be perceived as "excessive use, dependency, and interference with auton-omy—all images of drug abuse" (Slomka, 1992: 16), a similar perception to that of the use of steroids by athletes. Other users of propranolol could include surgeons in the operating theater, students before a big exam, and soldiers before battle. Although it is likely that some regulation of the use of nootropics might emerge, particularly in professions in which they disrupt the natural levels of competition, in our competitive environ-ment and drug-oriented society, it is expected that use of nootropics will be commonplace.

The New Generation of Pharmaceuticals

Rapid developments in drug intervention for the brain continue. At the 1996 meeting of the Society for Neuroscience, it was reported that the drug ampakine CX-156 restored memory in elderly men to the condition of that of men in their twenties. It was announced that it would soon be tested on Alzheimer's patients. In a related development, a recent study found that estrogen may relieve symptoms of Alzheimer's disease in postmenopausal women (Hotz, 1997), thus extending potential hormonal treatment directly into the brain.

Similarly, the development of new clot-busting drugs such as tPA (tissue-type plasminogen activator) promises significant benefits for many of the half-million stroke victims who die or suffer permanent damage each year. If administered within three hours of ischemic stroke, it clears vessels and quickly restores blood flow to the brain. After that point, however, tPA can actually make the stroke worse (National Institute of Neurological Disorders and Stroke, 1995). Although tPA cannot be used on hemorrhagic strokes because it aggravates brain bleeding, in combination with brain-imaging technology, this and other drugs promise considerable benefits to many persons.

In addition to a new generation of pharmaceuticals, more direct delivery methods are being tested. In 1997 a catheter was implanted in the brain of a patient with Lou Gehrig's disease to deliver the drug GDNF (glial-derived neurotropic factor) by circumventing the blood/brain barrier ("Device Implanted in Man's Brain," 1997).

VIRTUAL REALITY AND THE BRAIN

The next decade is certain to see astounding advances in the fabrication of worlds of illusion through virtual reality (VR) technologies. Already initial marketing of rather crude VR equipment is under way as part of the ongoing computerization of social interactions. Basically, VR is a computer-mediated, multisensory experience designed to trick our senses and put us in an alternative reality. Current VR systems allow one or more participants to become actively immersed in a tailored environment. Although movies, computer games, and even radio have long provided crude "virtual" realities directed toward one or several senses, VR combines all the elements of other media forms and ultimately promises to extend this synthetic environment to all of the senses. As such, completely artificial worlds can be created that are sufficient to make the user feel that the artificial world that they appear to inhabit is real. Moreover, unlike movies and computer games in which the viewer/player must relate to a viewpoint or character that participates in the action,

[i]n immersive virtual reality, the senses are provided with naturalistic data in a form which requires little learning or interpretation.

The experience provided to the user appears far more immediate than other forms of media and consequently may have more impact. (Foster and Meech, 1995: 210)

VR research itself might be invaluable in provided models for clarifying perceptual and sensory-motor systems in the human brain (England, 1995; Findlay and Newell, 1995). The combination of VR, artificial intelligence, cognitive research, and neuroscience research will offer novel research armaments for studying brain function. Most of the recent initiatives for the development of VR have focused on recreational uses—including the possibilities of virtual sex—but medical, industrial, and, of course, military applications are extensive. VR systems for mapping the brain are already being implanted. The next decade promises vast improvements in VR resolution and fidelity and in reduction in the sensory lag time that now constrains its potential for therapeutic purposes. In the near future, however, VR is expected to have widespread use in treating a variety of mental disorders.

Despite considerable enthusiasm over the prospects for VR, there has been little consideration as to the possible dangers inherent in placing individuals into alternative realities. Although proponents argue that it is the person who voluntarily enters VR and retains control, in effect VR gives the computer complete control over input into the human senses, thus altering experience, emotion, and ultimately thought. As VR advances to include more senses and a more complete perceptual field, the more immersed an individual may become in the virtual reality and the harder it may become for some persons to distinguish the real world from the artificial one that can be manipulated to achieve easy pleasures.

At present, little is understood as to the extent to which VR may interfere with normal psychological processes, but it is possible that such interference may put at risk individuals in mental or emotional peril. Furthermore, even persons with minor neuroses or perceptual problems may find that their sensations and reactions are exaggerated in VR. "Their residual memories and learning may even become distorted upon returning to the real world" (Cartwright, 1994: 24).

The proposed uses of VR to treat mental illness therefore may have significant risks that require urgent analysis before diffusion of the techniques. Like electroconvulsive shock treatment and frontal lobotomies in prior decades, VR also raises ethical questions and concerns over behavioral control abuses. Even in the absence of such abuses, however, treatment by VR raises questions. Though the first targets of medical researchers will focus on treatments for phobias and fears, they are likely to quickly be extended to a wider range of mental and psychological disorders.

One problem is that once a patient is in VR, the motivation to change may be reduced or eliminated. Because VR will allow us to experience what we find comforting and provide a ready means to avoid anxiety and distress, it may become an effective means of social withdrawal, particularly for those who find reality threatening. Moreover, a patient who chooses to remain in VR for prolonged periods is not accessible to needed appraisal. Effective therapy encourages personal development and fulfillment, allows individual responsibilities to be defined and accepted, and assists individuals to achieve greater independence integrated with society's expectations and norms. According to Whalley (1993: 283), VR devices do not well prepare a patient to address such an agenda and in fact may be counterproductive. In the extreme would be situations in which VR is so attractive to a patient (e.g., a quadriplegic) that he/she becomes overly depressed when forced to return to reality or in which VR itself becomes an addiction.

VR might also adversely affect persons who are not considered at risk because it contradicts or short-circuits basic psychological principles that are taken for granted in the real world but that do not exist or operate differently in the virtual world. For instance, while reality contact is often used as an indicator of mental health and psychological adjustment to the world around us, VR represents a deliberate manipulation of the senses to create a hallucinatory state, thus producing a "very fine line" between some kinds of VR experiences and certain schizophrenia-type states (Cartwright, 1994: 24). VR also produces an environment in which persons can create parallel lives, often with a completely different set of physical, social, and emotional attributes. Ironically, although VR is potentially a formidable communication medium, it could lead to unprecedented isolation of individuals (Burdea and Coiffet, 1994: 254).

Other areas where VR confuses psychological principles are the deliberate creation of altered states of reality, disembodiment and rematerialization into a virtual body, and the prospect of electronically projecting one's ego-center into a virtual space beyond the real body. Although this de-centering of self may be illuminating, it could also be "destabilizing and destructive" (Cartwright, 1994: 25). Recent evidence of problems raised by immersion of some persons in the Internet should raise warning signs regarding VR, which promises to be an even more seductive experience. Though VR could be a positive substitute to drugs in certain pathological cases, it also could create a "whole new class of depersonalized addicts" (Burdea and Coiffet, 1994: 254). Substantial research into the potential problems of diffusion of VR technology should receive high priority in order to set public policy in this powerful intervention technique. Moreover, the potential use of VR for political purposes and for abuse warrants closer examination, given the history of other technological extensions into the brain.

POLICY ISSUES ON BRAIN INTERVENTION

The arsenal of techniques and strategies for intervention in the brain are expanding rapidly and will be joined in the future by even more remarkable capabilities. Although many physicians and their patients welcome these broadened opportunities for treatment of mental disorders, their advent is not without concern. This concern is directed not only at the potential risks and unanticipated side effects that accompany any interventions in such a sensitive and complex organ as the brain, but also at the potential abuses of the powers of control that they bring. While new generations of psychoactive drugs and invasive procedures as well as completely new technologies such as VR are introduced and diffused with enthusiasm, the lessons of our past experience with cruder forms of intervention in the brain should not be ignored.

A large part of the legacy of earlier interventions in the brain, such as ECT and frontal lobotomies, is the fact that what is carried out voluntarily on some groups in society can easily be abused when used with coercion on other groups. Because of the inherent stigmatization of the mentally ill in U.S. society, caution is warranted when treatment is offered to the most vulnerable of groups.

Also, because these interventions can be used by society and the medical profession to control or alter behavior as well as to treat disorders as defined, at times subjectively, the motivations behind each application must be scrutinized. This is especially true as we develop a greater capacity to intervene in the genetic bases of neural functioning and learn more about how to alter behavior through genetic means. This issue of consent has been raised throughout this book. More in-depth discussion, however, is appropriate here concerning issues surrounding the right to refuse treatment and the use of psychotropic drugs on children.

The Right to Refuse Treatment

A major policy issue with all medical treatment in recent decades has centered on the concept of informed consent. This notion assumes that the person contemplating treatment must be apprised of its risks and benefits and that he or she understand its implications. It also assumes that the patient is capable of making his or her decision free from constraints, both personal and social, imposed upon them. Explicit in these conditions is the assumption that the patient is mentally competent to understand what he or she is consenting to and what the consequences of the decision are.

As discussed earlier, informed consent is especially problematic when dealing with brain intervention, in part because of the experimental nature of using even well-tested techniques on an individual patient's brain and the irreversible nature of some of these techniques. More important, in cases of severe brain disorders or injury can informed consent be possible? In other words, because consent must come from the damaged organ, the target of treatment, questions arise as to what stage of brain damage is severe enough to preclude informed consent and, at that stage, how to ensure the best interest of such patients in the absence of consent.

There are few problems if the patient agrees to the recommended treatment, even when his or her competency is questionable. Normally, this requires substantiating consent from a proxy decision maker who serves as a guardian to the patient. Similarly, in cases in which a clearly competent person refuses treatment and poses no risk to himself or others, the decision not to treat should be clear-cut. The more difficult problems arise when a person of questionable competency, because of his or her mental state or neurological disorder, refuses treatment that others deem to be in the patient's best interest or when a competent but potentially dangerous person refuses treatment. This latter situation often occurs in cases of diagnosed mentally ill patients who refuse to take medication needed to stabilize them.

Several trends merge to frame the context for the issue of the right to refuse treatment. First, the advent of effective psychotropic drugs, in combination with attempts by society to save money, have led to the deinstitutionalization of large numbers of psychiatric hospital patients. In 1955 it was estimated that 560,000 persons were institutionalized in psychiatric facilities. By 1989 this figure had been reduced to near 125,000, a pattern that continues. Whereas several were moved into community-based programs, many others were basically forced to reenter what often became a hostile world. Although this movement reduced ritual abuse of overprescription, forced medication, and behavioral control tactics to induce docility common to many institutions, it created new problems in the community.

Another trend in the United States has been a right-to-refuse-treatment movement that arose in part as a reaction to involuntary treatment abuses. According to Deaton and Bursztajn (1994), this movement has been successful in the regulation of antipsychotic medication and has clarified procedural protections of patients. In fact, they argue that it has been so successful that it has negatively affected the appropriate use of antipsychotic medications (1994: 97). It has also fueled controversy and raised potential conflict between the public fear of the mentally ill, especially where crime is involved, and the rights of the mentally ill patients to refuse treatment.

The debate over what pressures can be put on a person to undergo treatment to restore normality and perhaps protect others is complicated, and often the proponents of the various positions talk past each other. What is normalcy, and should stability of the patient be preeminent? Although in some cases medication may work to retain or restore personal identity, in others it may not. Some proponents of patients' rights argue that it is the patient, not society, who must make the decision as to whether normalcy is the preferred state of mind.

The history of attempts to limit the power of the government to impose treatment on unwilling patients is a recent one and has involved a variety of legislative and judicial actions. Because of their traditional responsibility in health-related policy, state legislatures have been active in setting statutory limits on involuntary treatment of mental patients. California Penal Code (Sec. 2670-2680), for instance, declares that all persons "have a fundamental right against enforced interference with their thought process, states of mind, and patterns of mentation." No drug, ECT, ESB, or other treatment that inflicts pain can be used without a court order on persons incapable of consenting. The authorities must then prove to the court that the therapy will be beneficial, that its administration is supported by a compelling state interest, and that no less intrusive therapy is available. A Florida statute (Sec. 394.459(3)(a) 1995) requires "express and informed consent" for treatment and sets careful procedural criteria for the authorization of involuntary treatment that is essential to appropriate care. A voluntary patient who refuses consent to treatment or revokes consent must be discharged within three days (Winock, 1997: 120). Other sources of state limits include judge-based tort law and regulatory limits imposed by state administrative agencies. The result of these actions is that many states have granted civilly committed patients the right to refuse treatment with antipsychotic medication (see Blackburn, 1990).

In most cases, states allow exceptions only after due process procedural reviews have taken place. However, if it can be demonstrated that a mentally ill patient endangers himself or others, lower courts have been willing to permit involuntary treatment with assessments carried out in the context of formal judicial hearings and full patient right of counsel (see *Rennie v. Klein*, 1983; *Souder v. McGuire*, 1976; *Davis v. Hubbard*, 1980).

The court's interest in the right to refuse treatment generally has emerged from First Amendment protections against forced government intrusion into the mind by altering the patient's thought processes. In *Kaimowitz v. Michigan Department of Mental Health* (1973), a Michigan circuit court found that the involuntary use of psychosurgery under the sexual psychopath law violated the First Amendment because it would irreversibly impair "the power to generate ideas." Even though Kaimowitz had signed the consent form, the court found that imposition of an involuntarily detained prisoner was unconstitutional.

In *Rogers v. Okin* (1979), a district court found a right to refuse psychotropic drugs, which it argued had the potential to affect the patient's capacity to think. Similarly, in *Bee v. Greaves* (1984), the U.S. Court of Appeals for the Tenth Circuit recognized the right of pretrial detainees to refuse administration of antipsychotic drugs. The court found that the drugs can affect the "ability to think and communicate," thus violating First Amendment protections. The U.S. Court of Appeals for the Seventh Circuit in *Lojuk v. Quandt* (1983) extended this thinking to ECT in ruling that the imposition of this procedure against the wishes of a voluntary patient in a V.A. hospital violated his right to refuse.

To date, the U.S. Supreme Court involvement in right-to-refuse-treatment cases has been limited to prisoners. In *Washington v. Harper* (1990), the Court held that the government cannot use psychotropic drugs to silence or punish prisoners without due process. Likewise, in *Riggins v. Nevada* (1992), the Court ruled against medicating prisoners to make them competent to stand trial. It did, however, in both cases frame its argument as a "limited liberty" interest rather than as the stricter fundamental constitutional right to refuse, finding that where a criminal's dangerousness is attributable to mental disorder, involuntary administration of antipsychotic drugs may be permissible without a judicial hearing when adequate alternative procedures are in place (Shapiro, 1994: 187).

In interpreting the actions of state legislatures and the courts regarding the right to refuse treatment, it is important to look at the setting. Although the Supreme Court ruled against involuntary administration of antipsychotic drugs for punishment in a prison setting, they have not agreed to hear similar cases from other types of institutions or the community. Also, in neither case was there a question of safety to the person or to other inmates or staff. By and large, the courts have recognized the heightened degree of state interests in institutional settings and the responsibility of the institution for safety. Therefore, the courts have been likely to defer to the institutional officials as long as procedural protections are met.

Clearly, the right to refuse treatment for mental disorders is relative, not absolute. It is dependent on the setting, duration, extent, and effects of the treatment, as well as on the competency of the intended subject. The lines between involuntary and voluntary intervention are not clear. There are many different degrees of coercion possible, just as each patient poses different degrees of potential danger to himself and others. Moreover, the procedures vary greatly as to safety, efficacy, and intrusiveness. It seems logical that the right to refuse a highly experimental, risky, and irreversible procedure such as psychosurgery be stronger than the right to refuse psychotherapy.

Winock agrees that the intrusiveness of the intended intervention is critical to decisions on the right to refuse treatment and that a certain threshold of intrusiveness may be posited before we consider a constitutional right to refuse treatment. He offers an admittedly rough

> continuum of intrusiveness based on the extent to which each treatment causes pain, produces harmful side effects, causes irreversible damage, invades bodily or psychological privacy, involves degrading procedures, and produces anxiety, fear, anger, or other negative reactions. (Winock, 1997: 25)

The resulting continuum (Table 6.4) from most to least intrusive is not mutually exclusive, since categories contain many components of varying intrusiveness depending on the circumstances under which they are applied. Along with the variability in setting, degree of consent, and motivation, this continuum demonstrates why the right to refuse treatment is such a difficult policy issue. Add to this the fact that in some instances there are very compelling issues of potential harm to others, and we have a very volatile rights issue.

Psychotropic Drugs and Children

An issue of growing concern centers on the potential for overmedication of children. Although this issue extends to all prescribed drugs, including

TABLE 6.4
Intrusiveness of Treatment Techniques

(Most intrusive)
    Psychosurgery
    Electronic stimulation
    Electroconvulsive therapy
    Psychotropic drugs
    Behavior therapy
    Psychotherapy
    Counseling
(Least intrusive)

Source: Adapted from Winock, 1997: 26.

antibiotics, antiasthmatics, and anticonvulsants, attention has recently focused on the use of psychotropic drugs. Because the vast majority of drug trials are conducted only on adults, approximately 80 percent of drugs are not approved specifically for children. However, once FDA approval for sale is given, physicians can prescribe it to anyone for any purpose. For psychotropic drugs in particular, this is highly problematic because of the lack of any data about the possible long-term effects on developing brains. An additional worry is that antidepressants, unlike antibiotics and other drugs, are often taken for many years, and proper dosage is difficult to calculate, especially for younger children.

As noted earlier in discussing Ritalin use, drugs prescribed to treat behavioral disorders and problems are especially controversial when given to large numbers of children. The prescription of Ritalin to over 1.5 million children daily, in addition to many others with a prescription of the stimulant Dexedrine, raises questions concerning the accurate diagnosis of attention deficit hyperactive disorder (ADHD), which is thought to affect considerably fewer children. Even for children who are accurately diagnosed with ADHD, some observers argue that unknown long-term consequences, slowed height development, and other side effects warrant caution in the tendency to treat with Ritalin when in doubt.

Most recently, attention on potential overmedication of children has focused on the rapidly escalating prescription of antidepressants, particularly Prozac. In one year, between 1995 and 1996, the use of SSRIs including Prozac, Zoloft, and Paxil by children increased by 60 percent. In 1996 nearly 600,000 children and adolescents were prescribed one of these drugs. Interestingly, this growth corresponded to the saturation of the adult market, and there has been a decrease in adult prescriptions over the past two years. It also came at a time when insurance companies were turning to drugs as a less costly alternative to conventional psychotherapy. As with most other drugs, none of the SSRIs have been tested and cleared for use by nonadults, despite their widespread use in children.

It is estimated that at least 4 million children and adolescents suffer from depression and are thus candidates for use of Prozac or alternatives. Many millions more may also be targeted for use as SSRIs are prescribed for behavioral problems. The lack of sufficient data on the safety and efficacy of these drugs does not appear to be a detriment to their widespread use. More important, because these drugs treat symptoms rather than causes, it is questionable what benefit they have in the long run for many of these children. The prescription of drugs to treat serious psychological and emotional disorders in children fails to address the social dimensions. Also, for minors, the issue of consent is often confused. Although one might assume that most parents and guardians place the child's health as predominant in consenting to drug therapy, experiences with Ritalin suggest that at least in a minority of cases the prime reason for consent is to better manage or control the child's behavior. Though informed consent is always at issue when any type of intervention in the brain is involved, with children the issue is clouded even further.

CONCLUSIONS

Whether the intervention treatment utilizes chemical, electrical, surgical, or computer techniques, issues of safety, efficacy, and consent are inherent in any attempt to modify the structure and functioning of the brain. Moreover, broader social issues arise because often the interventions are directed at suppressing symptoms, not at treating the root cause of the disorder. This is especially problematic when the patients are children or other vulnerable persons who are unable to exercise full informed consent, or when the long-term effects of the intervention are unknown.

The discussion here demonstrates the urgent need for more systematic, anticipatory analysis of the social consequences of the rapid diffusion of these often dramatic innovations. Because of our heavy dependence on technological solutions to health and social problems, and the prominence of the medical model of health, it is difficult to curtail or slow the widespread use of the latest wonder drug or procedure. Active marketing and publicity often encourage use—for example, Prozac in the late 1990s and frontal lobotomies in the 1950s—before the risks of the intervention are fully understood. Thus, it is even more crucial that the policy issues raised here are thoroughly examined. Like basic genetic and neuroscience research, these intervention techniques have significant social implications that warrant analysis by policy and social scientists as well as the health community.

7

Neural Grafting

Over the past decade there has evolved considerable enthusiasm among medical researchers on procedures designed to treat neurological disorders by grafting foreign tissues into the brains of patients. Neural grafting was elevated to a national debate that focused on the utilization of neural cells from aborted human fetuses. Unfortunately, the debate fueled largely by antiabortion forces placed attention on only one aspect of a complex set of policy issues tied to the research and application of neural grafting. It therefore diverted attention from significant broader concerns over safety, efficacy, informed consent, and the allocation of medical resources.

In addition to serving as an excellent example of the impact of abortion politics on research, neural grafting represents a case of the strong pressures exerted in the United States to move a biomedical innovation rapidly from experimental to clinical status and thus to blur the experiment/therapy distinction. Moreover, because neural grafting is a very invasive technique on the brain, it raises questions of when a person with a severe neurological disorder is capable of informed consent. Finally, the neural grafting experience demonstrates the politics of biomedical research and the difficulties inherent in framing a workable national policy to deal with the range of issues it generates.

NEUROLOGICAL DISORDERS AND THE POTENTIAL OF NEURAL GRAFTING AS TREATMENT

As discussed in Chapter 1, neurological disorders are a significant cause of illness, disability, and death in the United States. In 1990 the Office of Technology Assessment (OTA) identified those disorders that might be amenable to treatment by neural grafting if the current speculative predictions of its utility prove correct. The OTA emphasized, however, that at best only a subset of the patients suffering from any particular disorder is likely to benefit from neural grafting (OTA, 1990: 93). Even so, the number of individuals who might be helped by this approach is staggering, as will be the costs should we move in that direction. Table 7.1 shows the prevalence of the disorders identified as amenable to neural grafting.

Neurological disorders can result from either injury or disease. Injury to the central nervous system (CNS) can result from physical damage to the brain or spinal cord (skull fracture, concussion, wounds, broken backs) or from disruption in the normal flow of blood to the brain. Similarly, because the CNS is fragile and complex, it is susceptible to a number of diseases. Neurodegenerative disorders are marked by the loss of specific nerve cell populations in the brain or spinal cord. In most cases, the disease is progressive and the cell loss is gradual. The nature of the functional loss or impairment associated with a particular neurodegenerative disorder is directly related to the population of neurons affected. Examples of neurodegenerative disorders include Parkinson's disease, Huntington's disease, amyotrophic lateral sclerosis, and Alzheimer's disease.

Demyelinating disorders are marked by a loss of the fatty material (myelin) that surrounds many axons in the brain and spinal cord. When a cell loses this myelin sheath, its ability to send messages is impaired. The most common demyelinating disorder is multiple sclerosis (MS). MS is a progressive destruction of the myelin sheath, which normally acts to speed the conduction of electrical impulses. When the sheath is destroyed,

TABLE 7.1
Prevalence of Neurological Disorders in the United States

Neurological disorder	Prevalence
Alzheimer's disease	1 to 5 million
Stroke	2.8 million
Epilepsy	1.5 million
Parkinson's disease	500,000 to 650,000
Multiple sclerosis	250,000
Spinal cord injury	180,000
Brain injury	70,000 to 90,000
Huntington's disease	25,000
Amyotrophic lateral sclerosis	15,000

Source: OTA, 1990: 93.

the conduction of electrical nerve impulses is retarded, leading to muscle weakness, numbness, loss of coordination and balance, dizziness, slurred speech, fatigue, and loss of bowel and bladder control. Presently, there is no cure and therapy is of limited effectiveness.

The final form of neurological disorder is epilepsy, or the disruption of the normal electrical activity of the brain. Epilepsy can involve either a specific confined area of the brain or the entire brain and is manifested by episodes of abnormal electrical activity or seizures. Because the mechanisms underlying each disorder are different, they require varied roles for any treatment, including grafting.

Neurological dysfunction can be the result of injury as well as disease. The major cause of death and disability among children and young adults in the United States is brain injury through accidents or violent assault (U.S.D.H.H.S., 1989). Although the consequences of brain injury depend on the amount and type of damage sustained, the death of the damaged cells cannot be reversed. The basis for neural grafting for brain injury treatment, therefore, requires the replacement of the degenerated nerve cells.

Just as with the brain, the magnitude of functional loss (i.e., paralysis, loss of sensation) caused by spinal cord injury is directly related to the extent of the damage and where the injury occurs in the spinal cord. Spinal cord injury can result from a sudden impact, the gradual compression of the cord tissue, or the interruption of the blood supply to the spinal cord. Several strategies for the use of neural grafts to repair or substitute for the damaged neural pathways to induce recovery of function—including the stimulation of regrowth, the bridging of the injured region, or the replacement of nerve cells in the spinal cord—have been proposed for testing in animals (OTA, 1990: 101).

Stroke is the third leading cause of premature death in the United States after heart disease and cancer. A stroke is a sudden interruption of the normal blood flow to the brain, caused by the blockage or rupture of a blood vessel. The resulting shortage of oxygen or nutrients reaching the brain can produce a wide range of neurological deficits, including coma, paralysis, or death. As with other brain injury, damage from stroke cannot be reversed, though some animal experiments are under way that investigate the possibility of replacing the dead nerve cells with neural grafts.

Although neural grafting is in its most primitive stages, the review of these disorders shows that it holds promise of new therapeutic interventions in a wide range of conditions. The possible uses of neural grafting into the brain and spinal cord are varied, diverse, and far-reaching. Significantly more animal research is essential in order to begin to separate potential from fact, but the preliminary enthusiasm for this area of research is understandable.

Neural Grafting Procedures

Neural grafting will be employed to accomplish different treatment goals in different neuropathological disorders. There are at least three possible functions of neural grafting. One approach would focus on the provision of a continuous supply of chemical substances that have been depleted by injury or disease in affected regions of the brain or spinal cord. A second function would introduce new substances or cells that promote neuron survival, stimulate neuron regrowth, or both. The third approach would replace nerve cells in the brain or spinal cord that are lost to injury or disease.

In order to accomplish these varied functions, appropriate materials for transplantation must be selected. Sources for grafting include:

1. Peripheral nerve cells
2. Peripheral autonomic tissues
3. Tissues from outside the nervous system
4. Isolated, cultured, or genetically engineered cells
5. Fetal CNS tissue

Although all of these sources are likely to be used under some circumstances, two types have been the center of human research into Parkinson's disease. Prominent among the tissues from outside the nervous system are cells taken from the adrenal gland, which closely resemble neurons. Particularly applicable are the adrenal medullary cells, which are located in the innermost region of the gland.

Many scientists consider the most effective material for neural grafts to be human fetal CNS tissue (Freed et al., 1992; Lindvall et al., 1990). Fetal tissue is especially well suited to grafting because it replicates rapidly and differentiates into functioning mature cells. Unlike mature CNS tissue, fetal tissue has been found to readily develop and integrate into the host organism. Nutritional support provided by blood vessels from the host is readily accepted and likely promoted by fetal tissue. In animal experiments, fetal CNS tissue has displayed a considerable capacity for survival within the CNS of the graft recipient. Moreover, of all the graft materials available, fetal tissue is most capable of reconstituting nerve cell structure and function. Furthermore, fetal CNS tissue is amenable to cell culture and storage via cryopreservation and enjoys immunological advantages over other sources of material (OTA, 1990; 43).

In 1988 the first announcements of the grafting of fetal CNS tissue into the brains of Parkinson's patients were made by Mexican and Swedish research groups (Madrazo et al., 1988; Lindvall et al., 1988). Other countries in which clinical trials with fetal CNS tissue were begun include Great Britain, China, Spain, Czechoslovakia, Cuba, and, through private funding on a limited basis, the United States. After fifteen months, the first American patient to undergo the procedure showed improvement in movement and motor coordination (Freed et al., 1990). Forty-six months after the procedure the patient was doing well as were six patients with later fetal tissue grafts. These results led Freed et al. to conclude that "[f]etal tissue implants appear to offer long-term clinical benefit to some patients with advanced Parkinson's disease" (1992: 1549). Spencer and associates also found that while patients continued to be disabled by their disease, the symptoms and signs of parkinsonism were diminished during the eighteen months of evaluation (1992: 1541).

Despite case reports of success, many researchers remain skeptical of fetal CNS tissue grafting. For instance, Stein and Glasier argue that extensive primate studies have not yet been done. This means that patients are being subjected to procedures for which the long-range prognosis is unknown. If the recent revelations on the damaging long-term effects of breast implants contain a message, it is that we must examine the long-term effects of intrusive procedures before considering their routine use (1995: 367).

Similarly, Fahn concludes that the surgical procedure remains investigational and that reports with longer follow-up are essential (1992: 1590). He also asks whether the results justify the effort and cost of such a complex procedure. The rush to proceed with human trials without having collected adequate data from animal experiments is premature although understandable, in light of the pressures from desperate patients—and their physicians—who must deal with the realities of a devastating illness and are willing to undergo almost anything that might help them. Many persons in the medical/scientific community feel that additional basic animal research, coupled with very limited human experimentation, is preferable to the proliferation of human applications. According to Gash and Sladek,

> too many questions remain unanswered about the use of embryonic nerve cells to propose anything more than limited fetal grafting

in humans. . . . Although scientifically it seems logical to proceed, considerable information is needed before therapeutic success may be predicted. (1989: 366)

Reasons for concern over the unbridled rush to move to clinical application of fetal CNS grafts center on a number of potential risks as well as policy concerns over informed consent, the blurring of the line between experimentation and therapy, and the politics of abortion. As with any invasive surgical intervention, neural grafting presents risks to the patient. Immune system rejection of transplanted materials is possible. Although CNS allografts do not appear to suffer immediate rejection, long-term rejection is possible. If fetal tissue grafts are open to rejection, immunosuppressants are likely to be utilized, thus raising risks common to other transplantation procedures.

Assessment of adrenal medulla grafts in Parkinson's patients demonstrates that some psychological effects consistently accompany the procedure. These unwanted changes include hallucinations, confusion, and somnolence. In addition to psychological alterations, the more invasive graft placements into the brain can cause serious physical damage, including excessive bleeding or injury to the brain tissue. Injury from the surgery might result in the loss of CNS function or might exacerbate the recipient's immune system response to the grafted tissue. Furthermore, surgery temporarily disrupts the protective blood/brain barrier near the graft site, posing further threat to the patient's health.

Another risk to the patient is the possibility of excessive growth of the graft material. Fetal CNS tissue, some nonneuronal tissue, and continuous cell lines can continue to replicate in the CNS, thus presenting a risk of excessive graft enlargement. This expanding mass of tissue might itself compress and permanently damage the host brain tissue; obstruct the flow of cerebrospinal fluid, thus dangerously increasing pressure on the brain; and pose an added threat of tumor formation (OTA, 1990: 53).

Finally, transplanted cells can transmit bacterial and viral infections, thus placing the recipient at risk for diseases such as hepatitis, AIDS, or herpes. Yale scientists have discovered that one out of five fetuses harvested for transplant is infected. Moreover, in some experimental animals the implantation of fetal CNS tissue has been found to produce seizures (Gage and Buzsaki, 1989).

Another important caveat is that research to date has focused on the ability of grafts to provide relief from symptoms and to reduce disability. It has not addressed the capacity of neural grafts to provide cessation or reversal of the degenerative process. In fact, it is possible that the graft might be susceptible to the same pathological processes that underlie the disorder being treated. In other words, even if a neural graft takes and reduces the patient's symptoms, at some point the grafted tissue might succumb to the same degenerative process that affected the original cells. If so, the whole enterprise is of questionable value in treating Parkinson's disease.

The importance of informed consent is heightened in light of the risks the grafting patient faces. Does a desperate patient with a neuro-degenerating disease have the capacity to exercise free, informed consent when the object of the disease is in the brain? This has contributed to a dilemma in the testing protocol itself. In most instances, the subjects of clinical trials in neural grafting have been in the later stages of Parkinson's disease, where the risks of the procedure become secondary to the hope for any improvement at all. It might be that the most opportune time for grafting efficacy would be in the earliest stages of the disease, when the subject is relatively healthy—the very time at which the risks might be perceived as too great. Unfortunately, the intense policy focus on the fetal tissue debate over the past five years has diverted attention from the questions of informed consent of the subject and the relative risks and benefits that neural grafting presents.

THE POLICY CONTEXT

The major policy issue surrounding neural grafting using fetal CNS tissue has centered neither on the protection of experimental subjects nor on funding priorities, but rather on the controversy over the use of fetal tissue in transplantation research. Four possible sources of fetal tissue include that produced by:

1. Spontaneous abortions
2. Induced abortions on unintended pregnancies
3. Induced abortions on fetuses conceived specifically for research or therapy
4. Embryos produced in vitro

A dependence on spontaneously aborted fetuses for research is impractical because of the limited number available, the inability to control the timing of the abortion, and the fact that fetal tissue is fragile and deteriorates quickly after the death of the fetus (Garry et al., 1992: 1595). Moreover, because only tissue from fetuses in the eighth and ninth weeks of gestation can be used in neural grafting procedures, it may take as many as eighty human fetuses to yield enough tissue for one transplant. The major supply of fetal tissue, therefore, is likely to come from induced abortions on unintended pregnancies. Instead of discarding or destroying the tissue, it can be retrieved for research or transplantation.

The million and a half elective abortions performed in the United States each year would appear to be more than adequate to meet present research and transplantation needs. However, if neural grafting proves effective for patients with Alzheimer's disease, this supply would likely be insufficient. Based on the percentage of abortions performed between weeks six and twelve, the percentage performed by the safest method, and the fraction of fetuses in which the fetal midbrain can be identified, each year approximately 90,000 fetuses could be available for transplantation of brain cells (Fine, 1988: 6). Unlike the use of fetuses spontaneously aborted, though, dependence on elective abortions as the primary source of transplant material raises vehement objections on moral grounds by groups opposed to abortion. Other serious questions are raised when the research/therapy needs affect the timing and method of abortion, or if pressures are placed on the woman undergoing the abortion to consent to donation of her fetus.

An even more troublesome source of fetal material is found in those situations in which a human fetus is conceived specifically for the purpose of aborting it for research or therapy. Several cases have arisen in which individual women have attempted to produce tissue or organs for a relative or friend. In one case, the daughter of an Alzheimer's patient asked to be inseminated with the sperm of her father and, at the appropriate stage, to have the fetus aborted to provide her father with fetal neural tissue for transplantation (Krimsky, Hubbard, and Gracey, 1988: 9). Although there is no evidence at present that this is technically possible and the women's request was denied, this case demonstrates the potential demand for such applications. Similarly, another woman requested that her mid-term fetus be aborted and the kidneys be transplanted to her husband, who was dying of end-stage renal disease.

Whether the women in these two specific cases were making fully free decisions of their own, situations are bound to develop in which implicit or explicit coercion is placed by family members upon the woman to undergo this process. There is also concern that increased pressures for fetal tissue could lead to a marketplace for this scarce resource, which in turn could lead to the exploitation of poor women (in the United States or elsewhere), paid to conceive solely to provide fetal materials.

The final source of human tissue is embryos produced through in vitro fertilization. Again, the situation is clouded by the need to distinguish between those embryos deliberately created for research purposes and those embryos not transferred to the uterus after in vitro fertilization of multiple ova ("spare" embryos that would otherwise be destroyed or frozen indefinitely). Although persons who believe that life starts at conception are likely to oppose any use of human embryos for research purposes, there are many others who find the production of human embryos specifically for research unacceptable. Questions of consent, ownership, and payment are common to both categories, however.

Federal Regulations

In 1985, Congress passed a law (42 U.S.C. 289) forbidding federal conduct or funding of research on viable ex utero fetuses, with an exception for therapeutic research or research that poses no added risk of suffering, injury, or death to the fetus *and* that leads to important knowledge unobtainable by other means. Research on living fetuses in utero is still permitted, but federal regulations require the standard of risk to be the same for fetuses to be aborted as for fetuses that will be carried to term.

In 1985 Congress also passed legislation (42 U.S.C. 275) creating a Biomedical Ethics Board, whose first order of business was to be fetal research. In 1988 Congress suspended the power of the secretary of the Department of Health and Human Services to authorize waivers in cases of great need and great potential benefit until the Biomedical Ethics Advisory Committee conducted a study of the nature, advisability, and implications of exercising any waiver of the risk provisions of existing federal regulations. However, in 1989 the activities of the committee were suspended, thereby leaving the question of waivers unresolved.

Federal regulations on fetal research, then, appear to be quite clear in allowing funding within stated boundaries. However, in two key areas—embryo research and fetal tissue transplantation research—federal funding was in effect prohibited. The de facto moratorium on embryo research existed between 1980 and 1995. One of the provisions of the 1976 regulations prohibits federal funding involving the embryo entailing more than minimal risk unless an Ethics Advisory Board (EAB) recommends a waiver on grounds of scientific importance. The EAB was chartered for this purpose in 1977, was first convened in 1978, and in May 1979 recommended approval for federal funding of a study of spare, untransferred embryos. However, in September 1980 Health, Education and Welfare secretary Patricia Harris allowed the charter of the EAB to expire. Although some observers have speculated that Harris did so to avoid overlap with the planned presidential commission, Fletcher concludes that she instead did so out of opposition to federal funding of in vitro fertilization (IVF) research. According to Fletcher (1993: 212), Harris was fully aware that the EAB was the only lawful body that could recommend a waiver of minimal risk in research.

Although National Institutes of Health (NIH) directors throughout the 1980s called for a recharter of the EAB, no HHS secretary took action until 1988. Under pressure from Congress, Robert Windom, assistant secretary for health, announced that a new charter was to be drafted and a new EAB appointed. The draft charter was published in the *Federal Register* as required by law, and the charter was approved by HHS secretary Otis Bowen shortly before he left office. The incoming Bush Administration, however, never acted on it, and the EAB was not reestablished. As a result, the moratorium on federal funding of all human embryo research, including IVF and other assisted reproduction techniques, continued. According to the Institute of Medicine, this moratorium severely hampered progress in medically assisted reproduction. In 1994 the Human Embryo Research Panel, set up by the NIH, recommended federal funding of certain types of embryo research.

State Regulations of Fetal Research

As with federal activity in fetal research, state regulation was largely a response to the broad expansion of research involving legally aborted fetuses after *Roe v. Wade* (410 U.S.113 [1973]). Many state laws were enacted by conservative legislatures as an effort to foreclose social benefits that might be viewed as lending support to abortion. Of the laws specifically regulating fetal research in twenty-five states, twelve apply only to research with fetuses prior or subsequent to an elective abortion, and most of the statutes are either part of or attached to abortion legislation. Moreover, of the thirteen laws that apply to fetuses more generally, five impose more stringent restrictions on fetal research in conjunction with an elective abortion.

Under state law, research on fetal cadavers is regulated through the Uniform Anatomical Gift Act (UAGA), which has been adopted by all fifty states. However, some states have excluded fetuses from the UAGA provisions and others regulate those provisions through fetal research statutes. Although a total of forty-five states permit the use of tissue from elective abortions, fourteen have provisions regulating research involving fetal cadaver tissue that deviate from the UAGA either in consent requirements or in specific prohibitions on the uses of such tissue. Five states currently prohibit any research with fetal cadavers except for pathological examinations or autopsies. Of these laws, four apply exclusively to electively aborted fetuses.

State laws regulating research on live fetuses (in utero or ex utero) generally constrain research that is not therapeutic to the fetus itself. Because these state laws were adopted in the context of the abortion debate, the primary focus is on research performed on the ex utero fetus. Twenty states regulate research on ex utero fetuses, while fourteen regulate research on in utero fetuses. Although the specifics of the prohibitions and the sanctions designated differ by state, most appear to prohibit research involving transplantation and nontherapeutic research.

Another restriction imposed by some of the fetal research statutes addresses concerns over remuneration for fetal materials or participation in research. At present, at least sixteen states prohibit the sale of fetal tissue, seven for any purpose and nine specifically for research purposes. Importantly, some of these prohibitions apply only to elective abortions, not spontaneous abortions or ectopic pregnancies. In some states, the penalties for violation are stiff. For instance, selling a viable fetus for research in Wyoming is punishable by a fine of not less that $10,000 and imprisonment of one to fourteen years.

THE FETAL TISSUE TRANSPLANTATION CONTROVERSY

Although fetal research has been embroiled in controversy for two decades, the recent debate has focused on the use of fetal tissue for transplantation research. On March 22, 1988, in response to a growing political controversy surrounding the publicity over the Mexican and Swedish attempts at grafting fetal CNS tissue into Parkinson's patients, Robert Windom, Assistant Secretary for Department of Health and Human Services, imposed a moratorium on federal support for fetal tissue transplantation applications, pending a report from a special panel he had directed NIH to convene to answer ethical and legal questions posed by a proposal for transplanting fetal brain tissue into patients with Parkinson's disease. A similar project that involved implanting fetal pancreatic tissue in patients with juvenile diabetes had previously been funded. However, the principal investigator decided not to proceed with the human studies until the committee's report.

In September 1988 the Human Fetal Tissue Transplantation Research Panel recommended that funding be restored (NIH, 1988a). By a 17-to-4 vote, the panel concluded that such research is "acceptable public policy." In order to protect the interests of the various parties, the panel recommended guidelines to prohibit financial inducements to women; prohibit the sale of fetal tissue; prevent directed donations of fetal tissue to relatives; separate the decision to abort and the decision to donate; and require consent of the woman and nonobjection of the father. Although the prolife lobby accused the panel of being stacked in favor of the research community, there was considerable consensus across the various interests represented on the panel concerning the appropriateness of fetal tissue transplantation research. The panel was chaired by retired federal judge Arlin Adams, who is a known opponent of abortion but open to discussion of the use of fetal tissue. Membership on the panel, whose nomination required White House approval, included persons committed to a prolife, antiabortion philosophy that includes opposition to the use of any tissue obtained from what they view as an immoral act (Culliton, 1988b: 1593).

In December 1988 the NIH Director's Advisory Committee unanimously approved the special panel's report without change and recommended that the moratorium be lifted (NIH, 1988b). The committee concluded that existing procedures governing human research and organ donation are sufficient to regulate fetal tissue transplantation. In January 1989 James Wyngaarden, the director of NIH, concurred with the position of the Advisory Committee and transmitted the final report to the assistant secretary for health (Wyngaarden, 1989). The report languished in the department without action until November 1989, when Secretary Louis Sullivan—under pressure from Senator Gordon Humphrey (R-NH), Congressman William Dannemeyer (R-CA), and other prolife congressmen—announced, in direct conflict with the recommendations, an indefinite extension of the moratorium. The secretary concluded:

> I am persuaded that one must accept the likelihood that permitting the human fetal research at issue will increase the incidence of abortion across the country. Providing the additional rationalization of directly advancing the cause of human therapeutics cannot help but

tilt some already vulnerable women toward a decision to have an abortion. (Sullivan, 1989)

This extension of the ban on fetal research funding was challenged by Congressman Ted Weiss (D-NY) on technical grounds. He warned that making the moratorium permanent could be construed legally as a rule and should thus be made subject to the formal rule-making process rather than simply announced by the administration (Hilts, 1990).

In April 1990 the U.S. House of Representatives Committee on Energy and Commerce Subcommittee on Health and the Environment held hearings on human fetal tissue transplantation research. In July 1991 the House passed HR 2507, which would have limited the authority of executive branch officials to ban federal funds for areas of research without the support of an ethics advisory board. In May 1992 the House voted to approve the language of the conference committee, which included privacy and consent provisions added by the Senate, and sent the bill on to the president for a certain veto. In anticipation of a likely override of his veto, Bush issued an executive order directing the National Institutes of Health to establish a fetal tissue bank from spontaneously aborted fetuses and ectopic pregnancies, even though there was little evidence that such a bank could provide sufficient amounts of high-quality tissue for transplantation (Vawter, 1993: 82).

The political controversy continued in June 1992, when a compromise bill was introduced in both houses. This bill in effect gave the administration one year to demonstrate that the tissue bank would work. If researchers were then unable to obtain suitable tissue from the bank within two weeks of a request, they could obtain tissue from other sources, including elective abortions. In October 1992 a Senate filibuster by opponents of fetal research cut off attempts at the bill's passage, leading Majority Leader Mitchell to vow that the bill would be the first order of business when the Senate reconvened in January 1993. On his second day in office, President Clinton, as expected, issued an executive order removing the ban on the use of fetal tissue for transplantation.

POLICY ISSUES UNIQUE TO FETAL TISSUE TRANSPLANTATION

Despite the president's opening the way for public funding of research involving fetal tissue for transplantation, many policy issues remain. Although many of these issues raise purely procedural questions, others require substantive policy analysis regarding social values and priorities. These issues include:

1. The determination of fetal death;
2. The distinction between viable and nonviable fetuses as tissue sources;
3. The suitability of federal and state regulations and guidelines on the use of fetal tissue;
4. Quality control of fetal tissue, including screening of tissue and storage procedures;
5. Issues surrounding the distribution system, including financial arrangements between physicians performing abortions and researchers using tissues, payment for cell lines, and the legality of designating recipients by the tissue donors;
6. Issues surrounding the procurement of fetal tissue, specifically, who consents, the procedures for consent, the timing of consent, what information must be disclosed, and who seeks consent; and
7. Questions as to permissibility of altering routine abortion methods or timing in the interest of increasing the yield of fetal tissue suitable for research or transplantation.

These issues are bound to intensify as the amount of transplantation research increases and the need for fetal tissue expands, particularly issues 6 and 7, which are discussed in more detail in the following sections.

Consent for Use of Fetal Tissue

The question of individual rights is clearly present in the debate over procurement of fetal tissue. Does a woman who decides to abort her fetus maintain any interests in the disposal of the fetal cadaver? In those states that include fetal cadavers under UAGA, proxy consent is required. The most logical proxy is the pregnant woman. Not only is she the next of kin, but the privacy argument suggests the woman's right to control her own body and its products. Observers argue that to deny the woman the opportunity to veto the use of fetal remains for transplant research or therapy denies her autonomy and that consent, therefore, must be obtained to protect the interests of the woman. Following this approach, the NIH fetal tissue transplantation research panel concluded that maternal consent is essential prior to the use of fetal cadavers for research.

In contrast, others argue that through abortion the woman abdicates responsibility for the fetus and that, as a result, there is no basis for seeking her consent concerning the disposition of the fetus. Once abortion has taken place, the tissue is no longer part of the woman's body; therefore, her claim to the use of tissues from her body carries little weight. Furthermore, since the woman clearly does not intend to protect the fetus, it is inappropriate for her to act as a proxy.

Another argument against requiring maternal consent is that such a requirement may be an unwelcome intrusion upon the woman already facing the torturous decision of abortion. For instance, in *Margaret S. v. Edwards* (794 F.2d 994 [5th cir. 1986], 1004) the U.S. Fifth Circuit struck down a Louisiana regulation that required the woman's consent for the disposal of the dead fetus. The court noted that informing a woman of the burial or cremation of the fetus intimidated pregnant women from exercising their constitutional right to abortion or created unjustified anxiety in the exercise of that right. For the sake of reducing a woman's emotional burden and preventing harm to her, the court ruled that the woman need not be informed of, nor consent to, the disposal of the fetus.

One might ask whether informing a woman that her aborted fetus may be donated for research is potentially unwelcome information. Although some women might find it an intrusion, others may welcome the opportunity to specify donation for that purpose. The dilemma is that while UAGA and state statutes require the woman's informed consent in order for her fetus to be a legal donation, *Margaret S.* suggests that this requirement may intrude upon her abortion decision and thus be unconstitutional.

Robinson (1993) raises an interesting issue surrounding the woman's decision regarding abortion once she has consented to in utero research on the fetus. The issue is whether the woman should be able to change her mind about the planned abortion once the experiment on the fetus has begun. If yes, the woman's autonomy is ensured at the cost of potential severe harm to the fetus. If no, limits on her autonomy are accepted in order to prevent possible harm to the fetus. Robinson argues that, in this situation, the woman should not be permitted to change her mind on the abortion once she has begun participation in the experiment. The woman freely chooses to limit her own autonomy when she agrees to in utero experimentation, and she should be clearly told that she is making an irrevocable decision. As Robinson notes, this is what happens in any abortion when the point of no return arrives—this point simply comes earlier in the process, when the experiment starts. In either case, even if the fetus were to survive, it would have likely suffered great harm needlessly. Whether one agrees with Robinson's conclusion or not, this scenario reiterates the need for full and informed maternal consent prior to any experimentation on the fetus.

Modifications of Abortion Procedure

Another concern in fetal research—especially in grafting procedures, where the pressures to obtain tissue of the most appropriate gestational age and optimal condition for transplantation are strongest—is whether the pregnant woman's medical care can be altered in order to meet research purposes. Although some observers approve of modifications in the abortion procedure if they pose little risk for the woman and she is adequately informed of them and consents, no one has publicly advocated changing abortion procedures that entail a significant increase in the probability of harm or discomfort to the woman. At this stage there is general agreement that the means and timing of abortion should be based on the pregnant woman's medical needs and not on research needs. The fetal tissue transplantation research panel made this requirement a high priority.

The difficulty of ensuring cooperation from abortion clinics and obstetricians in making fetal tissue available for research, attributed in part to their reluctance to meet the additional time and resource requirements, however, has raised concerns for maintaining an adequate supply of fetal tissue. Moreover, the availability of RU-486 and other abortifacients in the near future might diminish the supply of usable fetal tissue at a time when demand might increase should fetal tissue prove to be a successful treatment for a common disease. Despite near consensus that fetal tissue procurement should not pose significant risk to the pregnant woman and that procedural protections must be in place, pressures for an expanding supply of usable fetal tissue demand vigilance to minimize abuses.

GENERIC ISSUES APPLIED TO NEURAL GRAFTING

Although informed consent is a problem in any surgical procedure, it is more problematic with any experimental therapy than with a proven treatment because of the uncertainty over efficacy and risks. At the least, the prospective subject must be made aware of how little is known about the risks and consequences of participation in the experiment. Because of the clinical setting in which neural grafting is conducted, usually under the auspices of a prestigious university medical center, the risks are especially difficult to convey. Moreover, it may be tempting for medical researchers to paint a more positive picture than is warranted because of their own high expectations, particularly when faced with a very desperate patient who may be perceived as having nothing to lose by the experimental procedure. Furthermore, it has been suggested that researchers might filter empirical evidence through their interpretations of the patient's current or future quality of life (Gervais, 1989), colored by their own enthusiasm for the potential benefits of the experimental procedure.

Although Stein and Glasier agree that neural grafting holds out the promise of potential treatment of a variety of disorders of the CNS, they contend that many problems must be resolved before the treatment can be routinely applied to patients (1995: 45). They argue that the medical, ethical, and legal risks of neural grafting should be fully assessed before proceeding with transplant surgery. Likewise, William Freed, a pioneer in neural grafting, concludes:

> An unfortunate emergence of excessive publicity about brain tissue transplantation has been the application of tissue transplantation as though it were a therapeutic procedure. Tissue transplantation remains an experimental technique and should be applied to humans only in the course of carefully planned therapeutic trials. (1990: 1434)

Freed is most disturbed by the use of neural grafting for diseases other than Parkinson's disease, especially for schizophrenia and Huntington's disease.

Furthermore, because of the blurred distinction between experimentation and treatment that is natural when a procedure is so rapidly moved from animal research to the human clinical setting, it has become difficult to distinguish a patient from a subject. Although we have institutional mechanisms to protect both the human subject and the patient (institutional review boards [IRBs] and hospital ethics committees, respectively), the procedures and criteria differ. As discussed earlier, the additional problem that neural grafts pose for informed consent is that a person is being asked for consent to intrude in the very organ that is dysfunctional enough to warrant experimental surgery. This concern is especially pertinent when dealing with disorders that affect mental functioning such as Alzheimer's disease. Federal regulations that require IRBs to add extra safeguards for subjects who suffer from acute or severe physical or mental illness should be rigorously applied in such cases.

Another issue at the center of neural grafting is whether experimental procedures should be regulated. According to the OTA, the

> intricate system of regulation to ensure the safety and efficacy of articles intended for use in the diagnosis, treatment, or prevention of disease in humans contrasts sharply with the absence of any direct Federal regulation of new surgical procedures developed for the same purposes. (1990: 127)

Although the FDA closely regulates the testing of drugs, for instance, it has not traditionally interfered in the practice of medicine. The new technologies involved in neural grafting, however, raise critical questions as to what constitutes the practice of medicine, under what circumstances the FDA should regulate what might be called the practice of medicine, and under what circumstances FDA jurisdiction to do so would be upheld (OTA, 1990: 125).

The final policy issue discussed here focuses on the broader question as to whether neural grafting ought to be a high social priority. As noted earlier, neural grafting embodies the technological fix approach to health care. The large amounts of money spent to possibly benefit each affected individual through neural grafting would be better spent on research to find the causes of these very devastating neurological conditions. Instead of hypothesizing about the potential to cure Alzheimer's patients through neural grafting—an approach that would in any case be prohibitive for the millions of individuals with the disease—neuroscience would do better to direct its scarce resources toward developing a more complete understanding of the disease etiology so that steps could be taken to avert or reduce the incidence of it where possible.

Unfortunately, in the current medical funding context, it is easier to obtain resources for high-cost clinical applications than it is for research directed at preventing future cases that require intervention. At present, no cost figures are available for neural grafting procedures should they become effective. Although this is understandable, given the current stage of development, provisions of even preliminary figures would demonstrate how prohibitive this "curative" approach is likely to be.

Already, despite the lack of anything approaching evidence of the efficacy of neural grafting, the expectations of the affected populations have been raised and demands for access to this procedure have been heightened. This demand, in turn, will increase pressures on the research community to proceed even more rapidly toward broader clinical trials, with an eye toward quick status as treatment. The lifting of the federal moratorium on the use of fetal tissue for transplantation will certainly accelerate the interest and the resources drawn to neural grafting, which inherently is an exciting and dramatic technology. The question as to whether society can afford this new endeavor, however, should be one for debate during the Decade of the Brain.

EXTENSIONS OF NEURAL GRAFTING

Recent research in related areas clearly demonstrates that the issues surrounding the use of fetal neural tissue for transplantation will intensify in coming decades. If the research is successful, the demand for fetal tissue will multiply. Fetal bone marrow transplants, for instance, show promise for treating such diverse diseases as leukemia, sickle cell anemia, and AIDS. In one study, bone marrow extracted from fetuses that had been miscarried were transplanted into baboons. The fetal bone marrow was twenty-three times more effective than adult marrow in terms of potential for self-renewal and proliferation rate, and it was significantly less likely to be rejected by the host.

In another study, fetal spinal cord cells were used to treat a patient with syringomyelia, a degenerative disorder that causes expanding holes to develop in the spinal cord. Fetal cells were injected into these cavities with the goal of having new nerve tissue grow to fill them and prevent further deterioration and damage to the cord. Although in this case the patient's treatment was not successful, if this research is successful, eventually spinal cord injury might be repairable. At this stage, however, slowing the disease is the best that can be promised. Another approach could be to harvest precursor nerve cells from fetuses and, through genetic engineering, the use of growth stimulants, or by other means, produce growing nerve cells in culture for transplantation to the central or peripheral nervous systems.

CONCLUSIONS

As noted in the introduction to this chapter, neural grafting cogently illustrates the complex policy issues underlying any invasive intervention in the brain. The highly emotional controversy over the use of fetal tissue from abortions, however, did little to illuminate the broader issue of experimentation on patients with neurological disorders. It also left unanswered an array of procedural and substantive issues regarding government responsibility in regulating a medical procedure with such significant public policy ramifications.

Furthermore, the debate failed to generate meaningful discussion as to whether this line of research and the development of yet another extremely costly procedure should demand high priority in public funding, as compared to research on the prevention of neurological disorders. Chapter 8 discusses the relative paucity of enthusiasm for prevention-oriented research as compared to that engendered by the dramatic medical intervention techniques represented by neural grafting.

8

Neurotoxicity

Nervous system dysfunction during advanced age seems destined to become the dominant disease entity of the twenty-first century. Neither I, nor anyone else, can tell you how much of that dysfunction might be attributed to toxic chemicals in the environment. So far, hardly anyone has looked. (Weiss, 1985)

One critical area of neuroscience that has received considerably less policy priority than is warranted concerns the role of neurotoxic substances in the development of neurological and some psychiatric disorders. Although every major body system can be adversely affected by toxic substances, the nervous system is particularly vulnerable. The range of potential neurotoxins is extensive, including pesticides; food additives; industrial chemicals; radioactive chemicals; microbial, plant, and animal toxins; cosmetic ingredients and pharmaceuticals; substance drugs, including alcohol and nicotine; foods; and even naturally occurring substances such as lead and mercury. The lack of high policy priority on neurotoxicity might result from the fact that the list of potential toxins is so inclusive and the task of identifying particularly hazardous ones is so daunting. Therefore, despite knowledge of the potential dangers of many substances, we still lack scientific proof for all but a handful.

The vulnerability of the nervous system to toxic substances is heightened because unlike other cells of the body, nerve cells cannot regenerate once lost and thus severe toxic damage to the brain or spinal cord is often permanent. Moreover, the dependence of the nervous system on a delicate electrochemical balance for normal functioning provides numerous opportunities for disruption from outside chemicals. These effects are accentuated because nerve cells, with their long axons, provide a vast surface area for chemical attack (OTA, 1990: 67).

152

Even minor changes in the structure or function of the nervous system can have profound consequences for neurological or behavioral functions. Also, because nerve cell loss and other regressive changes in the nervous system occur progressively with age, toxic damage may progress with aging and become more evident in the aging population. For instance, it has been found that exposure to neurotoxins early in life may not provide obvious symptoms until many years after exposure has ceased. One explanation of this latent damage is that at younger ages a brain is able to compensate for some adverse effects. As we age, however, the capacity of the brain to compensate diminishes and the damage inflicted much earlier becomes apparent (OTA, 1990: 70). In contrast, exposure to toxins in the second half of the prenatal period can be especially injurious to the rapidly developing nervous system, raising special concern for the fetal environment, an issue discussed later in this chapter.

In spite of the vulnerability of the nervous system to toxic substances that alter normal activity, knowledge of neurotoxicity remains surprisingly primitive. According to a recent report by the Office of Technology Assessment, "The number of substances that pose a significant risk to public health and the extent of the risk are unknown because the potential neurotoxicity of only a small number of chemicals has been evaluated effectively" (1990: 4). Few of the over 70,000 chemicals in the U.S. Environmental Protection Agency's (EPA) inventory of toxic chemicals have been tested to determine if they adversely affect the nervous system. Proof of neurotoxicity is very difficult for several reasons. Demonstrating causality is problematic, primarily because experimental human research is largely precluded in this area. As a result, evidence must come from two types of data, both with inherent constraints: retroactive epidemiological studies and animal experimentation.

Two forms of epidemiological studies are cohort studies and case-control studies. Cohort studies involve the comparison of different groups based on the composition of each group, whereas case-control studies compare the frequency of exposure to a suspected agent by persons with a particular disease to the frequency of exposure by comparable disease-free persons. Often epidemiological studies are initiated only after case reports establishing anecdotal connections between exposure and effect have alerted medical authorities to a potential problem and ecologic studies have uncovered raw correlations between the particular disorder and the frequency of exposure. Given the difficulty of prognosis of neurological disorders, even determination of the need for extended research is problematic.

Epidemiological studies must examine a large number of relevant factors, controlling, where possible, to isolate factors common to a particular disorder or to specific patterns of chemical exposure (e.g., exposure to chemical weapons in Iraq by U.S. troops who could be readily identified). The results in most every case are a set of statistical associations indicating increased risk related to exposure. In addition to the normal problems of inferring causality from aggregate measures of association, epidemiological studies on neurotoxic effects have special problems because the effect of most agents depends on many mediating factors, such as dosage, intensity of exposure, timing of exposure, and constitutional differences among individuals.

The second major source of data on neurotoxic effects comes from animal experimentation. Data collected from controlled animal experimentation are considered to be the best available in most cases. However, because of interspecies differences in effect and problems centering on the dosage and intensity of exposure, caution must always be used in extrapolating from animal data to humans. Nowhere is this more critical than in studies involving the brain and spinal cord, given the vast differences between humans and other species, including other primates. Although the presence of neurotoxic effects in other species requires examination in humans, the absence of neurotoxic effects in other species cannot be viewed as excluding such a possibility for humans.

In order to understand the effect of a toxin on the nervous system, the mechanism through which the substance interrupts normal functioning must be clarified. Some agents might act directly, whereas other agents might interrupt the processes indirectly, either by metabolic processing to a direct-acting toxicant or by altering the normal endocrine balance. Moreover, following exposure, the toxic agent must be distributed to the target tissue or organ where it exerts its toxic effect. Although the blood/brain barrier was at one time assumed to protect the brain, certain regions of the brain and nerves are directly exposed to chemicals in the blood and therefore to neurotoxic chemicals. Considerably more knowledge about the normal functioning of the brain is essential to understanding the precise mechanisms of action of potential neurotoxic substances. Even though we might be able to establish statistical relationships between exposure and adverse effects on the nervous system, at this stage it is difficult to determine how and why these effects occur.

A further complicating factor in assessing the mechanisms and effects of potential neurotoxins is that the impact of exposure to one agent may be altered by simultaneous exposure to other agents. Multiple exposures raise the problem not only of identifying the toxic agent responsible but also that of the interaction between agents in the expression of toxic effects. Combinations of agents may give rise to entirely new toxic effects not apparent in any of the agents alone. Therefore, studies that find that a substance is not toxic might be misleading by not accounting for its combined effect.

EFFECTS OF NEUROTOXINS

The lack of scientific evidence of the definite neurotoxic effects of many agents, however, should not minimize the deleterious impact of such substances and their potential costs to the mental health of the population (see Table 8.1). Although exposure to a few substances, such as trimethyl tin, can permanently damage the nervous system after a single exposure, in many more cases substances produce neurotoxic effects that are less dramatic and identifiable. Some substances may produce neurotoxic effects that are immediate but short-lived. For other substances, the adverse effects may appear only after repeated exposures over relatively long periods of time. Still others may lead to addiction, which represents a long-term adverse alteration of nervous system function. As noted above, in most instances the degree of damage will depend on many variables including the intensity and duration of exposure, and might be mediated by other factors including the genetic constitution of the individual. In some cases, substances that are safe and even necessary for the diet—such as vitamins A and B6 and iron—can cause neurotoxic reactions in large doses, particularly in young children. Also, chemical agents that alone are minimally toxic can be heightened in combination with other agents.

The first indications of damage to the nervous system might be subtle behavioral changes. An individual exposed to a neurotoxin may experience feelings of anxiety or nervousness, which may progress to depression, insomnia, memory loss, confusion, or speech impairment. In severe cases delirium and convulsions might appear, but often behavioral toxicological testing might be needed to detect an impairment.

Toxic substances can alter both the structure and function of cells. Structural alterations include changes in the morphology of the cells and the subcellular structures or the destruction of groups of cells. Toxins can affect the biochemistry and physiology of both neurons and glia. As a result of oxygen deprivation, the cells swell, the contents become more acidic, and their biochemical processes such as protein synthesis and

TABLE 8.1
Selected Major Neurotoxicity Incidents

Year	Location	Substance	Effects
1950s	Minamata, Japan	Mercury	46 deaths, infants born with nervous system damage
1950s	France	Organotin	Over 100 deaths
1950s	Morocco	Manganese	Severe neurobehavioral problems
1950s–1970s	United States	A ETT	Component of fragrances found to be neurotoxic
1956–1977	Japan	Cliquenol	Diarrhea drug found to cause neuropathy
1968	Japan	PCBs	1,655 affected when PCBs leaked into rice oil
1971	United States	Hexachlorophene	Infant disinfectant found to be neurotoxic
1973	Ohio	MnBk solvent	Exposure in fabric plant, 80 suffered polyneuropathy
1974–1975	Virginia	Kepone	Insecticide exposure, 20 suffered severe neurological problems
1976	Texas	Phosvel	Insecticide exposure, 4 suffered severe neurological problems
1977	California	Telone II	Pesticide exposure, 24 hospitalized
1979–1980	Texas	Lucel-7	Industrial exposure, 7 suffered serious neurological problems
1981	Spain	Toxic oil	20,000 poisoned by toxic substance in oil, many suffered severe neuropathy
1987	Canada	Domoicacid	Ingestion of mussels, 129 illnesses, 2 deaths

Source: Adapted from OTA, 1992: 47.

neurotransmitter secretion become inhibited. "At the morphological level, toxic substances seem to act selectively on the various components of the nervous system, damaging neuronal bodies (neuropathy), axons (axono-pathy), and myelin sheaths (myelinopathy)" (OTA, 1990: 67). As a result, neurotoxins often cause a slow degeneration of the nerve cell body or axon, leading to permanent neuronal damage. Illness stemming from exposure to neurotoxic substances often goes undetected because of the subtlety of neurotoxic responses and because their effects become apparent only over a long period of time, among other reasons.

Neurotoxic agents can cause a wide array of adverse functional consequences, from disruption of vision and lack of coordination, to memory loss and learning impairment, to modification of motor activities (see Table 8.2). In rare, severe cases it can lead to paralysis or death. In less severe cases the neurotoxic effects are reversible and diminish over time, with no lasting damage once exposure is terminated. In more severe cases, however, the effects are irreversible and lead to permanent damage in the nervous system since the nerve cells cannot normally regenerate.

TABLE 8.2
Neurological and Behavorial Effects of Exposure to Toxic Substances

Motor effects	*Sensory effects*
weakness	equilibrium changes
convulsions	vision disorders
tremor	pain disorders
lack of coordination, unsteadiness	tactile disorders
paralysis	auditory disorders
reflex abnormalities	
hyperactivity	*Cognitive effects*
	memory problems
Mood and personality effects	confusion
sleep disturbance	speech impairment
excitability	learning impairment
irritability	
restlessness	*General effects*
depression	loss of appetite
nervousness	depression of neuronal activity
delirium	narcosis, stupor
hallucinations	fatigue
	nerve damage

Source: OTA, 1992: 46

Although the deliberate development and use of highly toxic neurotoxins for terrorist activities (such as those used in the Japan subway) has received worldwide attention and concern, the cumulative impact of unplanned though less dramatic exposures to neurotoxins on a day-to-day basis undoubtedly is a more significant risk to populations in the twenty-first century.

POLICY ISSUE: THE FETUS IN THE WORKPLACE

One area of toxicity that did reach the political agenda during the 1980s centered on the impact of workplace hazards on the developing fetus. The public interest in workplace hazards was accompanied by heightened concerns over women's responsibility to fetal health during pregnancy, leading to policy measures designed to protect fetal interests. Women who used cocaine, and in some cases alcohol, during pregnancy were prosecuted, and other behavior such as smoking was condemned as contrary to the interest or the right of a fetus to be born with a sound mind and body (see Blank, 1992).

Unfortunately, considerably less attention was directed toward potential hazards external to the pregnant woman's behavior that are known to put the fetus at risk (see Figure 8.1). The issue became embroiled in the debate over the legality and morality of fetal protection policies, thus diverting attention from the potential neurotoxic effects on both the fetus and the adult in the workplace. It is important to note that the issue of workplace toxins was able to engender strong policy interest only through conflict over women's rights to work where they choose to work and protection of fetal health by prohibiting many women that choice. After that issue abated, workplace toxins moved again to the background.

The Risks of Fetal Damage

A variety of chemical challenges during the fetal period can lead to profound and irreversible abnormalities in brain development, although in many cases the effects are more subtle. "The developing brain exhibits specific and often narrow windows during which exposure to endocrine disrupters can produce permanent changes in its structure and functions" (Erice Statement, 1996). One problem in determining potential risk is that the effect of a specific agent will vary substantially depending on its timing and dosage during gestation. The same toxicant that is lethal or produces a congenital malformation at one stage in the pregnancy might

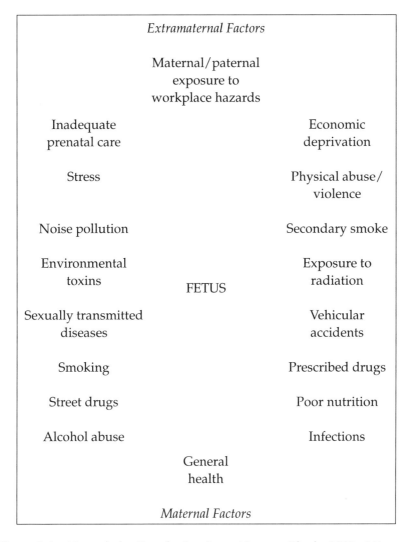

Figure 8.1   Hazards for Developing Fetus (*Source:* Blank, 1993: 12).

have no effect at another stage or dosage level. Prior to implantation, the pre-embryo apparently is resistant to certain toxicants. During this period of approximately two weeks a teratogen can be lethal, but should the embryo survive and implant, no demonstrable adverse effects are likely other than perhaps a development delay.

After implantation, the period of organogenesis is characterized by a high rate of cell division and by the timed differentiation of primordial cells into organ systems. This stage, lasting approximately from week two to week twelve in a pregnancy, is the most vulnerable period for induction of structural malformation (Elias and Annas, 1987: 198). However, because major functional and tissue maturation occurs throughout the stages of gestation illustrated in Table 8.3, an agent acting during this period can affect the overall growth of the fetus or size of specific organs and disrupt the developmental processes essential to fetal health. Also, interference with thyroid hormone function during a fetus's development can result in brain abnormalities. Moreover, because the brain continues to differentiate and the neural connections proliferate throughout gestation (though especially in the second and third fetal stages), exposure of the nervous system to toxicants may result in lasting behavioral deficits.

Another problem in establishing "safe" levels of exposure to potentially harmful agents during pregnancy is that tolerance varies substantially from one fetus to the next. The fetus is more susceptible than the mother to toxic environmental agents because of the high rate of cell division and the ongoing process of cell differentiation. Also, the fetus is often unable to metabolize and excrete a harmful substance and thus

TABLE 8.3
Stages of Fetal Development

Period	Weeks after conception	Stage	Days after conception
Embryo			
"Preembryo,"	first week	Zygote	1 to 2
"preimplantation		Cleavage	2 to 4
embryo," or		Blastocyst	4 to 6
"conceptus"		Implantation begins	7
Embryonic	2 to 3 weeks	Primitive streak	7 to 8
		Gastrula	7 to 8
		Neurula	20
	3 to 5 weeks	Limb buds	21 to 29
		Heart beat	21 to 29
		Tail-bud	21 to 29
		Complete embryo	35 to 37
	6 to 8 weeks	Body definition	42 to 56
Fetus	9 to 40 weeks	First fetal	56 to 70
		Second fetal	70 to 140
		Third fetal	140 to 280

Source: OTA, 1988; adapted from Blank, 1984.

bears its entire effect, whereas the mother may suffer no ill effects. As a result, it is difficult to frame a policy that responds adequately to both the woman's employment rights and the danger of harm to the fetus.

Out of the tens of thousands of chemicals and substances commonly present in workplaces, the one with the most well-documented links to fetal damage to the central nervous system is lead (U.S. Environmental Protection Agency, 1986). In pregnant women, lead rapidly crosses the placental barrier early in gestation, when the fetus is most vulnerable to its effects (Crocetti et al., 1990). Female workers exposed to high lead levels are at abnormally great risk for spontaneous abortion and stillbirths. Maternal exposure to elevated levels of lead during pregnancy has been correlated with mental retardation, intrauterine growth retardation, and neurological disorders. The children of such women are also susceptible to convulsions after birth. And these effects may result from even relatively low levels of exposure. A recent study found that the fetus may be adversely affected at blood lead concentrations well below the 25mg per deciliter threshold currently defined by the Centers for Disease Control and Prevention as the highest acceptable level for young children (Bellinger et al., 1987: 1042).

Inorganic and organic mercury, too, have been found to cross the placenta in animals and humans and cause damage to the central nervous system of the fetus, resulting in mental retardation, cerebral palsy, seizures, blindness, and heightened rates of stillbirths (Smith, 1977). All forms of mercury appear to be teratogens, capable of altering fetal growth and increasing the incidence of congenital malformations, chromosomal abnormalities, and biochemical changes in the human placenta.

Many other substances widely used in industry are known or suspected to have adverse effects on the reproductive systems of exposed workers. Chromosomal damage has been reported in workers exposed to the solvent benzene, an element in paint strippers, rubber cement, nylon, and detergents. Anesthetic gases produce miscarriages and birth defects in the progeny of both male and female operating room and dental personnel (Cohen, 1980). Exposure to pesticides and chlorinated hydrocarbons, used to manufacture dry cleaning fluid and other general solvents, causes serious fetal damage (Howard, 1981). Workers exposed to vinyl chloride risk severe impairment to their reproductive systems, and exposure has been tied to abnormal rates of miscarriage and chromosomal damage of fetuses.

More than 75 percent of the workers who make semiconductors and printed wiring circuit boards are women. These women are exposed on a daily basis, and often under conditions of inadequate ventilation, to an extensive array of hazardous agents, including solvents, acids, resins, adhesives, sealants, rubber, and plastics. Furthermore, soldering and welding operations result in the formation of toxic gases and may involve solders of diverse metal content, including cadmium and lead. Toluene, polychlorinated biphenyls, and other known teratogens are also present in a variety of combinations.

Fetal Protection and the Courts

To protect the fetus—and to protect themselves—a number of companies established fetal protection policies in the 1970s and 1980s that excluded all fertile women from positions in which workers might be exposed to high levels of potential toxicants. From a fetal development perspective, these policies were reasonable because the structural effects of many toxins are often most severe early in the pregnancy, when the woman is frequently unaware of her condition. However, by eliminating all fertile women instead of pregnant women alone, the policies excluded most women from better paying positions. Throughout the 1970s and 1980s, the courts grappled with the issue as women sued to have these exclusionary policies overturned. Although Title VII of the 1964 Civil Rights Act included a prohibition against discrimination in employment "because of" or "on the basis of" sex, court rulings were mixed as to whether such policies constituted disparate treatment or had a disparate effect on women.

In October 1978 Congress passed the Pregnancy Discrimination Act, which included the Pregnancy Discrimination Amendment to Title VII of the Civil Rights Act. This amendment left no doubt that the intent of Congress was to include discrimination on the basis of pregnancy as a clear case of sex discrimination. The act stated that "women affected by pregnancy, childbirth, or related medical conditions shall be treated the same for all employment-related purposes . . . as other persons not so affected" (U.S. Code Sec. 2000 e, k, 1982). However, though the Pregnancy Discrimination Act extended the scope of Title VII to the whole range of matters concerning the childbearing process, it gave virtually no consideration to the issue of fetal vulnerability to workplace hazards.

There are two defenses available to employers in Title VII challenges. The first, the Bona Fide Occupational Qualification (BFOQ), can be met only under two conditions: (1) the qualification invoked must be reasonably necessary to the essence of the employer's business, and (2) the employer must have a reasonable factual basis for believing that substantially all women would be unable to safely and efficiently perform the duties of the job. The second defense for making such employment distinctions is the judicially defined defense of "business necessity." Under this defense, the business purpose must be "sufficiently compelling" to override any discriminatory impact.

In 1988 the first of two conflicting decisions (one federal, one state) involving the fetal protection policy of Johnson Controls was made by the U.S. District Court for the Eastern District of Wisconsin (*International Union, UAW v. Johnson Controls Inc.*, 1989). Suit was brought by unions and employees, claiming that this policy violated Title VII. The trial court, however, granted summary judgment to Johnson Controls, thus denying the claimants' arguments of discrimination.

Upon reargument of the case, the U.S. Court of Appeals, Seventh Circuit, affirmed the trial court's ruling. After citing a long history of concern for employee health by Johnson Controls (previously Globe Union), the appellate court found that the fetal protection policy was reasonably necessary to industrial safety and, thus, should be recognized as a BFOQ protection against claims of sex discrimination. According to the court, available scientific data indicated that the risk of harm to the fetus as a result of lead is confined to fertile females (p. 889). Therefore, the employer's fetal protection policy was based upon real physical differences between men and women relating to childbearing capacity and was consistent with Title VII. The court also held that the union failed to show that less discriminatory alternatives would be equally effective in achieving Johnson Controls' purpose of protecting fetuses from substantial risk of harm (p. 901).

In a conflicting state ruling, the California Court of Appeals, Fourth District, reversed a superior court decision that found that Johnson Controls' fetal protection policy did not violate the California Fair Employment and Housing Commission Act (*Johnson Controls, Inc. v. California Fair Employment and Housing Commission*, 1990). The appellate court held that the fetal protection program "unquestionably discriminates against women" because only women are affected by its terms (p. 160). Although the court admitted that the dispute was "fraught with public policy considerations" and pitted state interests in protecting the health of employees and their families against state interests in safeguarding equal employment opportunities for women, it concluded that categorical discrimination against a subclass of women, such as all of childbearing capacity except those proven to be sterile, violated the California statute prohibiting discrimination on the basis of sex. Contrary to the U.S. Court of Appeals in *International Union*, the California court held that Johnson Controls could not defend its fetal protection program as a BFOQ.

On March 10, 1991, the U.S. Supreme Court issued its ruling on *United Automobile Workers v. Johnson Controls Inc.* on an appeal from the Seventh Circuit. In a unanimous 9–0 decision (with two concurring opinions), the Court reversed the decision of the Court of Appeals and remanded the case to the Seventh Circuit. The Court stated that the bias in the policy of Johnson Controls is obvious because fertile men, but not fertile women, are given a choice about whether they want to risk their reproductive health by holding a particular job (p. 7). The Court noted that since its grant of certiorari, the Sixth Circuit had reversed a district court's summary judgment for an employer that had excluded fertile female employees from foundry jobs involving exposure to specific concentrations of airborne lead. In *Grant v. General Motors Corp.* (1990), the Sixth Circuit stated, "We agree with the view of the dissenters in *Johnson Controls* that fetal protection policies perforce amount to sex discrimination, which cannot logically be recast as disparate impact and cannot be countenanced without proof that infertility is a BFOQ" (p. 1310). The Supreme Court's majority opinion, written by Justice Blackmun, agreed that Johnson's fetal protection policy created a facial classification and required a BFOQ defense.

After establishing the need for a BFOQ defense, the Court turned to the question of whether Johnson Controls' policy was one of those "certain instances" that come within a BFOQ exception. In citing its previous holding in *Dothard v. Rawlinson* (1977), wherein danger to the woman herself does not justify discrimination, the Court reiterated that in order to qualify as a BFOQ, a job qualification must related to the "essence" or "central mission" of the employer's business. Furthermore, within the Pregnancy Discrimination Act (PDA), Congress made it clear that the decision to become pregnant or to work while being either pregnant or capable of becoming pregnant was reserved for each individual woman to make for herself. The Court concluded that the language of both the BFOQ provisions and the PDA, as well as the legislative history and the case law, prohibit an employer from discriminating against a woman because of her capacity to become pregnant unless her reproductive potential prevents her from performing the duties of her job. The Court, therefore, had "no difficulty concluding that Johnson Controls cannot establish a BFOQ" (p. 17).

Although the Court was divided on whether any fetal protection policies could ever conceivably be justified under the BFOQ defense—in concurring opinions, Judges White, Rehnquist, Kennedy, and Scalia did not rule this out—all nine justices agreed that Johnson Controls' policy could not be so justified. No matter how sincere was Johnson Controls' fear of prenatal injury, it did not "begin to show that substantially all of its fertile women employees are incapable of doing their jobs" (p. 18).

Unfortunately, the debate over fetal protection policies diverted attention from the broader range of hazards that face both women and men in the workplace. The reduction of these hazards and the inclusion of comprehensive preventive health initiatives will do more to maximize fetal health than did exclusionary workplace policies in traditionally male-dominated industries. Guaranteed access to prenatal care, pregnancy leaves with job security, and expanded employment opportunities for women should be among the top priorities of the coming years if the goal of healthy children and the reduction of toxins in the workplace is to be realized.

NEUROTOXICITY AND HEALTH

With all the attention now directed toward neurogenetics, neural grafting, neuropharmacology, and a growing array of imaging techniques and innovative interventions, the impact of neurotoxins on health has generated considerably less research support or public awareness. Given the vast effects of environmental insults on the nervous system and the large potential long-term costs of the damage imposed, this relative negligence of neurotoxins is unacceptable. Williams notes that though protecting the human brain should be an environmental public priority, it is not. "Hit a child on the head, causing an intellectual disability, and this is seen as a public concern. Drive a car using leaded petrol causing intellectual disabilities in countless children, and that is seen as a medical problem" (1996: 5).

Although considerable research effort has been directed toward cancer as the main outcome of environmental change, low priority has been given to what are perhaps even more important effects on the human brain. As we discover more direct associations between environmental neurotoxins and neurodegenerating diseases and mental disorders, the incentives for spending a considerably larger proportion of our resources on this research should increase. But given our current value system, it will likely not receive the attention warranted.

In part, our low priority on neurotoxicology reflects a general low priority on public health approaches and prevention strategies. In the

United States, individual rights have been elevated to a status where they are absolutely dominant over collective interests (Lemco, 1994: 6). Americans, as a result, are hesitant or unwilling to sacrifice perceived individual needs for the collective good, especially that of future generations. This perspective is manifested in our willingness to spend unlimited amounts of resources to save the life of an individual but considerably less to invest in programs designed to reduce the mortality and morbidity of many others. This contrast between an immediate identifiable patient and large numbers of statistical persons in the future is much starker in the United States than it is in other Western democracies, where prevention, public health, and community-wide interests are given more priority in the allocation of resources.

A resulting and reinforcing factor is that American culture is predisposed toward progress through technological means. This has led to both an overdependence on medical technology to resolve our health problems and the belief that medicine is synonymous with health. These beliefs persist despite substantial evidence that health outcomes of populations are more highly correlated with the economic and social environment than with how much they spend on medicine. Most health gains have been the result not of an improvement in medical care but of an improvement in the standard of living and broad public health measures (Blank, 1997: 56–71). For Frum:

> the tragedies counted among America's most horrific health problems—from premature underweight babies to AIDS-infected drug addicts to the twelve-year-old gunshot victims ... are not really health problems any more than the plight of the homeless is a housing problem. (1995: 33)

In spite of this knowledge, it is far more comfortable to search for the quick technological fix of symptoms than to find the real causes and prevent them by altering individual behavior and social conditions.

We continue, therefore, to identify high-technology medicine with health care and to feel most satisfied with care that uses sophisticated, state-of-the-art techniques. In turn, this fascination and demand for an unending stream of medical innovations has become a major factor fueling runaway health care spending in the United States (Aaron, 1991). It also leads to the overuse of technologies that have iatrogenic effects (e.g., bypass surgeries that result in neurological damage). Most important, this unrealistic faith in and dependence on technological medicine diverts attention and valuable resources from disease prevention, health promotion, and nonmedical health strategies (Mechanic, 1994: 8).

This medical model is firmly based both in the public and in the medical community of largely specialists trained in the technological imperative, and it is constantly reinforced by the mass media. Announcements of new neural-grafting procedures, the discovery of genes for neurodegenerating diseases or mental disorders, and new psychotropic drugs make the headlines. In contrast, unless they are directly tied to deaths of identifiable persons, neurotoxins do not get this exposure. The neurosurgeons on popular TV shows like *ER* and *Chicago Hope* performing amazing life-saving procedures are the heroes, the pathbreakers. Not surprisingly, there are no similar shows featuring the work of epidemiologists or neurotoxicologists. Moreover, announcements of dramatic medical breakthroughs are commonplace on the evening news and are largely offered without in-depth analysis of their implications. Again, notably absent are less dramatic but more critical findings that could lead to the prevention of disease or the discovery of the causes of many neurological disorders.

9

Conclusions:
The Emergence of Brain Policy

As in other areas of biomedicine, new developments in neuroscience will emerge at an accelerating rate over the coming decades. Furthermore, because we are still at the primitive stages of understanding the nervous system, the speed of discovery in neuroscience is likely to outpace that of other areas of medical science. This book has demonstrated that despite the recency of knowledge in neuroscience, the technological fruits of the Decade of the Brain represent merely an impressive start to a more complete understanding of brain function and dysfunction, the genetic and neurological basis of human behavior, and the biochemical foundations of mental disorders. Less impressive to date have been our gains in knowledge of the impact of the environment on the brain.

BRAIN POLICY

This book has raised a broad range of policy issues that accompany our unfolding knowledge of the brain and the application of an expanding array of techniques for intervention in the brain. In coming years, these issues should elicit considerable research attention by social scientists and bioethicists. This analysis demonstrates that the brain is a new biomedical policy area that must be studied with the same rigor as other more traditional areas such as genetics. Although brain policy reflects problems generic to other biomedical areas, because the impact of the brain is so pervasive in all areas of human life and death, it raises unique challenges to the study of political behavior and of the human condition in general. Brain policy, therefore, represents a new conceptual area of study that warrants urgent attention by policy makers, policy analysts, and informed citizens.

Furthermore, it is argued here that because at some level the brain moderates all human thought and action, brain policy is interrelated to other areas of biomedicine. Clearly, the linkage between genetics and neuroscience is intractable. As we come to understand the mechanisms and role of neurotransmitters and receptors and their relationship to genes, policy concerns inherent in these seemingly disparate substantive areas will merge. The lines drawn to date between genetic policy and brain policy are most likely a reflection of the relative primitive state of knowledge about the brain—in other words, a temporary condition. Already, much genetic research is directed at the amelioration of mental disorders or deficits or at behavioral problems.

Moreover, as the discussion in Chapter 2 illustrated, our conception of the brain is deeply tied to the full range of death-related issues. Our knowledge about the brain and our ability to measure brain activity more precisely will strongly influence our very definition of death. Similarly, policy on organ transplantation is dependent on the concept of brain death, as are the issues concerning the use of the organs of anencephalic infants. Less directly, brain policy is interrelated with policy on workplace hazards and the fetus, on fetal research, and on the treatment of severely ill newborns with Down's syndrome or other brain-centered problems. Given these linkages, brain policy takes on even more significance as an important new area of biomedical policy.

THE BRAIN AND HEALTH POLICY

The current attention directed toward neuroscience has raised awareness of its importance to all aspects of human life, but it has not yet placed neuroscience high on the policy priorities in health care. This, however, will change. The wide array of new intervention capacities and the tremendous costs of CNS-related health care problems, along with the emergence of the view of the inseparability of the mental and physical dimensions of health, will demand considerably more attention by policy makers in the coming decades. The health benefits of more specific psychotropic drugs, neurogenetic treatments for mental disorders and neurodegenerating diseases, and heightened understanding of brain function in general are significant.

Unfortunately, the emergence of these new intervention techniques comes at a time when health care resources are becoming scarce and competition for funding is tight. Many interventions, particularly medications, are likely to be cost-effective and in fact might lead to cost savings. However, other emergent treatment strategies such as neural grafting and neurogenetic procedures will be very costly on a per case basis. Moreover, because total cost involves frequency of use as well as per case cost, even less expensive procedures can add significantly to overall health care spending when applied to large populations, for example, Alzheimer's patients.

To date, seldom have even rough estimates been given for prospective innovating procedures such as neural grafting or preimplantation gene therapy. Although this is understandable at this early stage of development, analysis of cumulative costs is crucial before they come into routine use. As noted in Chapter 1, though high cost alone is not a justifiable reason to block or slow diffusion of various neural intervention techniques, as with all medical innovations it must be a factor for consideration. Although the inclusion of economic considerations is alien to our current value system, complete analysis of brain intervention techniques is impossible without such data.

PREVENTING ENVIRONMENTAL DAMAGE TO THE BRAIN

One area that I believe has not received the attention it deserves centers on the impact of environmental influences on the brain. As discussed in Chapter 8, the effect of neurotoxins on health remains poorly understood, not because we are unaware of the risks, but because we have placed relatively low priority on prevention in general. Unlike the other subjects discussed here, environmental neurotoxins lack the more immediate, dramatic context found in curative interventions.

Despite this relative neglect, the biggest benefits for the health of the population are likely to come from attempts to prevent neural diseases. Instead of placing highest priority on developing new and costly therapies, in the long term we are better served by finding the causes of mental disorders and diseases linked to the environment and then developing policies to reduce their incidence. Though prevention runs counter to the individual-oriented medical model because it places emphasis on statistical as opposed to identifiable patients, any policy that continues to minimize the proximate causes of neurological disorders will fail to maximize health no matter how many innovative treatments are invented.

Although some observers correctly argue that preventive efforts interfere with individual privacy and require behavioral changes, in dealing with the brain any approach risks such problems. Inherent in any brain intervention, whether preventive or therapy, are issues of privacy, autonomy, and confidentiality.

In addition to placing higher priority on searching for environmental contributions to neurological disorders, we must incorporate our new knowledge of the brain/behavior association to health care in general. As we come to better understand the neurogenetic roots of risk-taking behavior, addiction, and aggression, we should use this knowledge to develop appropriate social strategies for the prevention of the massive problems these behaviors cause. In light of the high incidence of brain injury and trauma, more emphasis on accident prevention is warranted, particularly among young people, for whom the long-term economic and personal costs are staggering. Likewise, reduction of domestic abuse, handgun injuries, and violence must be integral elements in a broad-based strategy to lessen the need for dramatic methods of brain intervention to treat the results of such actions.

HEALTH OUTCOMES IMPACT STATEMENTS

The inability of medical technology assessment (MTA) to challenge the medical model should be no surprise in light of the support given it by decision makers and shaped by the American value system. Despite the best intentions and efforts of the assessors, it is virtually impossible to break out of the confines created by society. Under conventional MTA, the process places the burden of proof on those persons or groups who want to block new technologies and procedures, a virtually impossible task given the vagaries and complexities of medical science. Any reasonable doubts are resolved in favor of going ahead with the development and diffusion of the technology.

If we are serious about ensuring that medical innovations (and existing practices) are indeed safe, effective, and efficient in producing health, the burden of proof for demonstration must be shifted to the proponents. This would put biomedical procedures on a more equal ground with drugs and biologics. I agree with Callahan that in weighing the consequences of technologies we should assume the worst outcome and put the burden on the optimists to prove that assumption wrong. Moreover, the more intrusive and irreversible a technique, the more rigorous should be the scrutiny it receives.

To this end, I argue that we should consider the concept of a health-outcomes impact statement analogous to environmental impact statements. The purveyors of new biomedical technologies—particularly new intervention areas such as neural grafting and preimplantation genetics—would need to attain approval from a health-outcomes board before they could claim their status as clinical procedures and receive health care dollars. Without approval, the interventions would remain on a preclinical or experimental status. This would have the added benefit of clarifying the experiment/therapy distinction that has become confused in much of the research discussed throughout this book. Informed consent vagaries that are now commonplace would be clarified considerably for potential subjects/patients. More important, the burden of proof not only for safety and efficacy but also as to an intervention's contribution to health should be rigorous, thus denying approval to intervention areas that are judged insufficient by these criteria. Although the health-outcomes board could not prohibit the use of techniques or procedures, reimbursement through the health care budget would not proceed until preliminary approval was obtained. Furthermore, any patients who underwent the intervention would have to be informed as to its current status in terms of health outcomes.

Though this proposal is bound to be attacked as unduly slowing medical progress and as an unjustified interference with medical prerogative, it is not a revolutionary concept. Countries with global budgets in effect do this on a regular basis. Before new techniques or procedures are reimbursed, proponents must demonstrate that the innovation will do the job better than existing methods and/or do it in a more cost-effective way. Although execution of this approach is more difficult in the private-insurance U.S. system, private insurers should embrace a process that protects them from paying for unproven and unapproved techniques. The health-outcomes impact approach would go a step further, however, in requiring evidence of a clear contribution to population health in proportion to the cost of the innovation, thus addressing the concern of Lomas and Contandriopoulos (1994) that alternative, nonmedical pathways to health be considered on an equal basis when assessing medical technologies.

THE POLITICS OF NEUROSCIENCE

The political debate surrounding this emerging knowledge about the brain and new intervention techniques promises to be intense. These new discoveries are already challenging prevailing societal values, as they demonstrate that much of what each of us is can be reduced to the actions of neurons and neurotransmitters. In the very least, neuroscience findings require a reevaluation of democratic concepts of equality, individual autonomy, freedom, and responsibility. Neuroscience also undermines conventional perceptions of human nature and of the notion of the mind as a blank slate to be written on by experience.

As demonstrated in earlier chapters, controversy over the brain is nothing new. Until recently, however, intervention techniques have been highly intrusive and crude (e.g., frontal lobotomies), and knowledge of how the brain functions highly speculative. Critics, therefore, were able to readily exploit these factors in their efforts to establish policy constraints in their use. Also, the intrusive physical nature of early procedures provided extreme cases of abuse for opponents to focus on. Still, the use of these procedures was curtailed in the end less because of the success of the critics and more because minimally intrusive methods, particularly drugs, replaced the cruder forms.

Although the more explicit and sensational physical risks of brain intervention have been reduced because the new techniques are both more effective and subtle, they may in fact be more problematic. First, these procedures are more difficult targets for criticism because they are less easily sensationalized. Ice pick surgery and abuses of primitive forms of electrical treatment generated outrage in large part due to their crudeness. In contrast, widespread use of psychotropic drugs is generally perceived as a legitimate area of medical therapy and cases of abuse are likely to be minimized by the public.

Second, because they appear less intrusive, newer interventions more readily fit a value system dependent on technological solutions. Rather than being feared, psychotropic drugs are welcomed as solutions to a wide array of personal and social problems and are generally portrayed in glowing terms by the mass media. These drugs are attractive because under the cloak of medicine they promise happiness, health, weight loss, and performance enhancement.

Just as the "old" eugenics is being replaced by a more subtle new form based on genetic screening and diagnosis, so the older forms of mind control are giving way to more implicit, sophisticated methods such as drugs and virtually reality, which are often embraced even as they exert their influence. Fears of eugenics and behavior control are more easily manifested by involuntary sterilizations and coerced psychosurgery than by routine but questionable uses of genetic screening and medication. In fact, the control aspects of the latter methods are likely to be more significant because they are less intrusive and more readily subsumed under the medical model.

Moreover, as behavioral disorders are transformed into neurological disorders caused by biochemical problems, they naturally become the province of medicine. The subject of a proposed intervention to alter behavior becomes a patient who, of course, must be treated whether he or she agrees or not. To conclude, the diffusion of more sophisticated and seemingly less invasive intervention techniques will likely complicate, not eliminate, concerns over control, whether genetic or neurological.

Perceptions of Affected Individuals

A related issue has to do with the impact of these technologies on our perceptions of candidates for their use. As discussed in Chapter 1, the stigma attached to victims of mental disorders is clearly manifest in the United States. Whether this stigma is exacerbated or abated by new interventions depends less on the innovations themselves and more on how society perceives them. Knowledge of the genetic bases of mental disorders, for instance, might lead either to increased sympathy for those dealt a poor genetic hand or to rigid genetic screening programs designed to reduce the number of individuals with those traits in the population.

Similarly, our views of violent offenders or those who exhibit other antisocial behaviors will be affected by neurogenetic research that leads to the discovery of biological bases of such behavior. On the one hand, this evidence could produce an empathy with affected individuals and intensified efforts to help these persons cope with what are basically biochemical problems. Eventually, this could alter the way the justice system handles offenders, by shifting emphasis from punishment to prevention and treatment. On the other hand, as with conventional mental disorders, this knowledge could result in heightened involuntary neurotropic interventions for individuals identified at special risk based on genetic testing and/or brain mapping. Again, the response to this knowledge is political and social.

The same situation applies for neuroscience findings regarding sex differences, sexual orientation, and addictive personalities or behaviors. Though technological advances in neuroscience can provide the basis for better understanding of human variation and the acceptance of differences, alternately they can work to exaggerate inequalities and lead to repression of the most vulnerable members of society. How they affect society depends upon the motivations behind their use. While there is little evidence at present that developments in neuroscience are motivated by anything other than beneficence, ultimately the political climate will determine to what ends they are put.

We must guard against a dual system in which interventions are used voluntarily to improve the lives of the already well-off but involuntarily to control the behavior of the least well-off. Another tiered system to be avoided is one that denies needed treatment to the least well-off members of society. This is especially critical for those interventions without which a person is unable to function as a full member of society or for enhancement interventions. For the former, the person will remain disadvantaged. For the latter, there is a danger that those persons who already enjoy advantages in capabilities will have the means to extend their advantage, thus widening the gap. These issues in distribution and access become more important in a society in which mental capacity is of the highest importance.

THE BRAIN AND SOCIETY

The brain is a major resource for society. Both individually and collectively, the health of the brain and the maximization of its potential should be high priorities. Moreover, as we move into the twenty-first century, the increased emphasis on technical skills will mean that brain power will be of even greater importance. All evidence suggests that the era of the need for a large unskilled labor force is past and that individuals without technical or professional skills will be left behind. Furthermore, in the competitive global economy, countries unable or unwilling to cultivate a highly trained workforce will suffer economically. Resources will disproportionately go to those with the necessary abilities to function in a more mental, less physical framework.

Despite all the emphasis on technological interventions, in the end we must return to the social environment to maximize brain potential. Chapter 2 demonstrated that the development of the brain is constantly shaped by the outside world. The prenatal environment and family setting, especially in early childhood, are critical for proper maturation and stimulation of the brain.

Moreover, the sensitive chemical balance of the brain makes it susceptible to environmental hazards, including malnutrition, neurotoxins, violence, substance abuse, and accidents. Ironically, one set of environmental insults to which we need to be especially attentive in the future are the interventions reviewed in this book. Each time we intervene in the brain or attempt to treat it we risk unanticipated consequences that may cause harm to this delicate but powerful organ. The complexity of each human brain and its wiring means that iatrogenic, or treatment-caused, harm is always a possibility—one that must be minimized by intervening only with strong justification.

The framework of politics is constantly changing. New developments in science are a major source of this change. For example, the revolution in telecommunications promises to either democratize politics or threaten democracy, depending on one's perspective and how the technology unfolds. Likewise, discoveries about the brain and its relationship to behavior will alter how we view political behavior and thus how we actually practice politics.

As noted in Chapter 5, the importance of the brain for political behavior has implications at three levels of analysis: individual, intra-societal, and extrasocietal. To the extent that our new understanding of the brain clarifies the neurological bases of aggression and the behavior of leaders and followers, it has significant ramifications for the conduct of domestic as well as international policy making. At the individual level, neuroscience has the potential to challenge empirical theories, including rational-choice theories and rational-voter models. This in turn makes suspect any methodologies that exclude biological variables including the actions of neurotransmitters. Dependence on survey research and other self-reporting methods is especially questionable in light of this knowledge.

The issues raised in this book will be prominently reflected in public policy discussions, but they will also shape the way in which the dialogue is conducted. The brain has always been the source of all political thought and action, though it has often been taken for granted or treated as an "empty" organism in political theory. No longer will the theories based on such assumptions be viable in the wake of our emerging understanding of how the brain works. To this end, neuroscience is contributing to our altered view of the human condition that will reverberate throughout conventional theories of politics.

Court Cases

Bee v. Greaves, 744 F.2d 1394 (10th Cir. 1984)

Davis v. Hubbard, 506 F.Supp. 915, 933 (N.D. Ohio 1980)

Dothard v. Rawlinson, 433 U.S. 321 (1977)

Grant v. General Motors Corp., 908 F.2d 1303 (1990)

International Union, UAW v. Johnson Controls, 680 F.Supp. 309 (E.D. Wis. 1988), rehearing en banc, 886 F.Supp. 871 (7th Cir. 1989), reversed 89-1215 (U.S. Sup. Ct., March 20, 1991)

Johnson Controls v. California Fair Employment and Housing Commission, 267 Cal.Rptr. 158 (Cal. App. 4 Dist. 1990), cert. denied 1900 Cal. LEXIS 2107 (1990).

Kaimowitz v. Michigan Department of Mental Health, Civ. No. 73-19434-AW (Mich. Cir., July 10, 1973)

Lojuk v. Quandt, 706 F.2d 1456 (7th Cir. 1983)

Rennie v. Klein, 720 F.2d 266 (3d Cir. 1983)(en banc)

Riggins v. Nevada, 504 U.S. 127 (1992)

Rogers v. Okin, 478 F.Supp. 1342 (D. Mass. 1979)

Souder v. McGuire, 423 F.Supp. 830 (E.D. Pa. 1976)

Washington v. Harper, 494 U.S. 210 (1990)

Bibliography

Aaron, Henry J. 1991. *Serious and Unstable Condition: Financing America's Health Care*. Washington, D.C.: Brookings Institution.

Abrams, R. 1988. *Electroconvulsive Therapy*. Oxford: Oxford University Press.

Ad Hoc Committee of the Harvard Medical School to Examine the Definition of Death. 1968. "A Definition of Irreversible Coma." *Journal of the American Medical Association* 205: 337–40.

Alkon, Daniel L. 1992. *Memory's Voice: Deciphering the Brain-Mind Code*. New York: Harper and Row.

Allen, Garland E. 1989. "A Dangerous Form of Eugenics is Creeping Back into Science." *The Scientist* (February 9): 9–11.

Allen, L. S., and R. A. Gorski. 1991. "Sexual Dimorphism of the Anterior Commissure and Messa Intermedia of the Human Brain." *Journal of Comparative Neurology* 312: 97–104.

Andreasen, Nancy C. 1997. "Linking Mind and Brain in the Study of Mental Illness: A Project for a Scientific Psychopathology." *Science* 275: 1586–92.

Angarola, Robert T., and Alan G. Minsk. 1994. "Regulation of Psychostimulants: How Much Is Too Much?" In Harold I. Schwartz, ed., *Psychiatric Practice under Fire*. Washington, D.C.: American Psychiatric Press.

Annas, George J. 1989. "Who's Afraid of the Human Genome?" *Hastings Center Report* 19(4): 19–21.

Bailey, J. Michael, and Richard Pillard. 1991. "A Genetic Study of Male Sexual Orientation." *Archives of General Psychiatry* 48: 1089–96.

Bailey, J. Michael, Richard C. Pillard, Michael C. Neale and Yvonne Agyei. 1993. "A Genetic Study of Female Sexual Orientation." *Archives of General Psychiatry* 50: 217–23.

Barinaga, Marcia. 1997. "Visual System Provides Clues to How the Brain Works." *Science* 275: 1583–85.

———. 1997. "New Imaging Methods Provide a Better View into the Brain." *Science* 276: 1974–76.

Bellinger, D., A. Leviton, C. Waternaux, et al. 1987. "Longitudinal Analysis of Prenatal and Postnatal Lead Exposure and Early Cognitive Development." *New England Journal of Medicine* 17: 1037–43.

Besdine, Richard W. 1995. Testimony at Hearing before Special Committee on Aging, U.S. Senate, 104th Congress, 2nd Session, June 27, *Breakthroughs in Brain Research*. Washington, D.C.: GPO.

Billings, Paul. 1994. "Genetic Discrimination and Behavioral Genetics: The Analysis of Sexual Orientation." In Norio Fujiki and Darryl Macer, eds., *Intractable Neurological Disorders, Human Genome Research and Society*. Tsukaba, Japan: Eubios Ethics Institute.

Blackburn, C. E. 1990. "The 'Therapeutic Orgy' and the 'Right to Rot' Collide: The Right to Refuse Antipsychotic Drugs under State Law." *Houston Law Review* 27: 447–573.

Blank, Robert H. 1984. *Redefining Human Life: Reproductive Technologies and Social Policy*. Boulder: Westview Press.

———. 1992. *Mother and Fetus: Changing Notions of Maternal Responsibility*. Westport, CT: Greenwood Press.

———. 1993. *Fetal Protection in the Workplace*. New York: Columbia University Press.

———. 1997. *The Price of Life: The Future of American Health Care*. New York: Columbia University Press.

Blank, Robert H., Lynton K. Caldwell, Thomas C. Wiegele, and Raymond A. Zilinskas. 1988. "Toward Better Education in Biopolitics." *Issues in Science and Technology* 4(3): 51–54.

Bootzin, Richard R., Joan Ross Acocella, and Lauren B. Alloy. 1993. *Abnormal Psychology: Current Perspectives*. New York: McGraw-Hill.

Bower, B. 1996. "Gene Tied to Excitable Personality." *Science News* 149: 4.

Burdea, Grigori, and Philippe Coiffet. 1994. *Virtual Reality Technology*. New York: John Wiley.

Callahan, Daniel. 1990. *What Kind of Life: The Limits of Medical Progress*. New York: Simon and Schuster.

Carey, Gregory, and Irving I. Gottesman. 1996. "Genetics and Antisocial Behavior: Substance versus Sound Bytes." *Politics and the Life Sciences* 15(1): 88–90.

Cartwright, Glenn F. 1994. "Virtual or Real? The Mind in Cyberspace." *Futurist* (March–April): 22–26.

Chalmers, David. 1996. *The Conscious Mind: In Search of a Fundamental Theory*. Oxford: Oxford University Press.

Chang, L. W., and R. S. Dyer, eds. 1995. *Handbook of Neurotoxicology*. New York: Marcel Dekker.

Changeux, Jean-Pierre. 1997. *Neuronal Man: The Biology of the Mind*. Princeton: Princeton University Press.

Chorover, Stephen L. 1981. "Psychosurgery: A Neuropsychological Perspective." In Thomas A. Mappes and Jane S. Zembaty, eds. *Biomedical Ethics*. New York: McGraw Hill.

Churchland, Patricia A. 1986. *Neurophysiology: Toward a Unified Science of the Mind-Brain*. Cambridge: MIT Press.

Churchland, Paul M. 1995. *The Engine of Reason, the Seat of the Soul: A Philosophical Journey into the Brain*. Cambridge: MIT Press.

Cleghorn, J. M., R. B. Zipursky, and S. J. List. 1991. "Structural and Functional Brain Imaging in Schizophrenia." *Journal of Psychiatric Neuroscience* 16: 53–74.

Cloninger, C. Robert, Rolf Adolfsson, and Nenad M. Svrakic. 1996. "Mapping Genes for Human Personality." *Nature Genetics* 12: 3–4.

Coffin, R. S., and D. S. Latchman. 1996. "Herpes Simplex Virus-Based Vectors." In David S. Latchman, ed., *Genetic Manipulation of the Nervous System*. London: Academic Press.

Cohen, Ellis N. 1980. "Waste Anesthetic Gases and Reproductive Health of Operating Room Personnel." In Peter Infante and Marvin Legator, eds., *Proceedings of Workshop on Methodology for Assessing Reproductive Hazards in the Workplace*. Washington, D.C.: National Institute for Occupational Safety and Health.

Cohen, William S. 1995. Opening Statement at Hearing before Special Committee on Aging, U.S. Senate, 104th Congress, 2nd Session, June 27. *Breakthroughs in Brain Research*. Washington, D.C.: GPO.

Comings, David E. 1996. "Both Genes and Environment Play a Role in Antisocial Behavior." *Politics and the Life Sciences* 15(1): 84–85.

Concar, D., and A. Coughlin. 1993. "Is There Money in Lost Memories?" *New Scientist* 1869: 20–22.

Cook-Deegan, Robert M. 1994. "Ethical Issues Arising in the Search for Neurological Disease Genes." In Norio Fujiki and Darryl Macer, eds., *Intractable Neurological Disorders, Human Genome Research and Society*. Tsukaba, Japan: Eubios Ethics Institute.

Crick, Francis. 1966. *Of Molecules and Men*. Seattle: University of Washington Press.

———. 1979. "Thinking about the Brain." *Scientific American* 241: 219–33.

Crick, Francis, and Christof Koch. 1992. "The Problem of Consciousness." *Scientific American* (September): 153–59.

Crocetti, A. F., P. Mushak, and J. Schwartz. 1990. "Determination of Numbers of Lead Exposed Women of Childbearing Age and Pregnant Women." *Environmental Health Perspectives* 89: 121–24.

Cross, Peter J., and Barry J. Gurland. 1987. "The Epidemiology of Dementing Disorders." Contract Report for U.S. Congress Office of Technology Assessment. *Losing a Million Minds*. Springfield, VA: National Technical Information Service.

Damasio, Antonio R., and Hanna Damasio. 1992. "Brain and Language." *Scientific American* (September): 89–95.

Dean, N. and H. Morganthaler. 1992. *Smart Drugs and Nutrients*. Santa Cruz: B. and J. Publications.

Deaton, Rodney, and Harold Bursztajn. 1994. "Antipsychotic Medication: Regulation through the Right to Refuse." In Harold I. Schwartz, ed., *Psychiatric Practice under Fire*. Washington, D.C.: American Psychiatric Press.

Dennett, Daniel C. 1984. *Elbow Room*. Cambridge: MIT Press.

———. 1991. *Consciousness Explained*. Boston: Little, Brown.

"Device Implanted in Man's Brain to Control Disease." 1997. *Sarasota Herald-Tribune*, January 15, 5B.

Dicter, H. 1992. "The Stigmatization of Psychiatrists Who Work with Chronically Mentally Ill Persons." In P. J. Fink and A. Tasman, eds., *Stigma and Mental Illness*. Washington, D.C.: American Psychiatric Press.

Double, Richard. 1996. *Metaphilosophy and Free Will*. New York: Oxford University Press.

DuPont, Robert L. 1995. *The Selfish Brain: Learning from Addiction.* Washington, D.C.: American Psychiatric Press.

Durham, J. 1989. "Sources of Public Prejudice against Electroconvulsive Therapy." *Australia–New Zealand Journal of Psychiatry* 23: 453–460.

Elias, S., and George J. Annas. 1987. *Reproductive Genetics and the Law.* Chicago: Year Book Medical.

Emanuel, Linda L. 1995. "Reexamining Death: The Asymptotic Model and a Bounded Zone Definition." *Hastings Center Report* 25(4): 27–35.

England, Rupert. 1995. "Sensory-motor Systems in Virtual Manipulation." In Karen Carr and Rupert England, eds., *Simulated and Virtual Realities: Elements of Perception.* London: Taylor and Francis.

Erice Statement. 1996. Consensus Statement issued at Erice, Italy, 20 May.

Etzioni, Amitai. 1978. "Individual Will and Social Conditions." *Annals of the Association for the Advancement of Political and Social Science* 437: 62–73.

Fahn, Stanley. 1992. "Fetal-Tissue Transplants in Parkinson's Disease." *New England Journal of Medicine* 327 (22) 1589–90.

Farina, A., J. D. Fisher, and E. H. Fischer. 1992. "Societal Factors in the Problems Faced by Deinstitutionalized Psychiatric Patients." In P. J. Fink and A. Tasman, eds., *Stigma and Mental Illness.* Washington, D.C.: American Psychiatric Press.

Finch, Caleb E., and Rudolph E. Tanzi. 1997. "Genetics of Aging." *Science* 278: 407–11.

Findlay, John M., and Fiona N. Newell. 1995. "Perceptual Cues and Object Recognition." In Karen Carr and Rupert England, eds., *Simulated and Virtual Realities: Elements of Perception.* London: Taylor and Francis.

Fine, Alan. 1988. "The Ethics of Fetal Tissue Transplants." *Hastings Center Report* 18(3): 5–8.

Fischbach, Gerald D. 1992. "Mind and Brain." *Scientific American* (September): 48–57.

Fisher, L. J. and Fred H. Gage. 1993. "Grafting in the Mammalian Central Nervous System." *Physiological Reviews* 73(3): 583–616.

Fitzgerald, Faith T. 1994. "The Tyranny of Health." *New England Journal of Medicine* 331(3): 196–98.

Fletcher, John C. 1993. "Human Fetal and Embryo Research—Lysenkoism in Reverse. How and Why?" In Robert H. Blank and Andrea L. Bonnicksen, eds., *Emerging Issues in Biomedical Policy, Vol. II.* New York: Columbia University Press.

Forkosch, Joel Anton, H. Stephen Kaye, and Mitchell P. La Plante. 1996. "The Incidence of Traumatic Brain Injury in the United States." *Disability Statistics Abstract* 14: 1–4.

Foster, Deborah, and John F. Meech. 1995. "Social Dimensions of Virtual Reality." In Karen Carr and Rupert England, eds., *Simulated and Virtual Realities: Elements of Perception.* London: Taylor and Francis.

Fox, Renee C. and Judith P. Swazey. 1992. *Spare Parts: Organ Replacement in American Society.* New York: Oxford University Press.

Freed, Curt R., et al. 1992. "Survival of Implanted Fetal Dopamine Cells and Neurologic Improvement 12 to 46 Months after Transplantation for Parkinson's Disease." *New England Journal of Medicine* 329(5): 321–25.

Freed, William J. 1990. "Fetal Brain Grafts and Parkinson's Disease." *Science* 250. 1434–35.

Frum, David. 1995. "What to Do about Health Care." *Commentary* (June): 29–34.

Gage, Fred H. and G. Buzaki. 1989. "CNS Grafting: Potential Mechanisms of Action." In F. Sell, ed., *Neural Regeneration and Transplantation*. New York: Alan R. Liss.

Garry, Daniel J., Arthur L. Caplan, Dorothy E. Vawter, and Warren Kearney. 1992. "The Use of Fetal Tissue in Research For Parkinson's Disease." *New England Journal of Medicine* 327(22): 1591–95.

Gash, D. M. and J. R. Sladek. 1989. "Neural Transplantation: Problems and Prospects—Where Do We Go From Here?" *Mayo Clinic Proceedings* 64: 363–67.

Gershon, Elliott S., and Ronald O. Rieder. 1992. "Major Disorders of Mind and Brain." *Scientific American* (September): 127–33.

Gervais, Karen G. 1989. "Neural Transplants and Nerve Regeneration Technologies: Ethical Issues." Paper prepared for OTA.

Goleman, Daniel. 1996. "Brain Images Show Science of Addiction." *Sarasota Herald-Tribune* August 13, A1.

Goldman, David. 1996. "High Anxiety." *Science* 274: 1483.

Goldman-Rakic, Patricia S. 1992. "Working Memory and the Mind." *Scientific American* (September): 111–17.

Goodwin, F. K., and K. R. Jamison. 1990. *Manic-Depressive Illness*. New York: Oxford University Press.

Gowan, Frank H. 1991. "Cocaine Addiction: Psychology and Neurophysiology." *Science* 251: 1580–1585.

Green, William. 1986. "Depo-Provera, Castration, and the Probation of Rape Offenders: Statutory and Constitutional Issues." *University of Dayton Law Review* 12(1): 1–26.

Greenspan, Stanley I. 1997. *The Growth of the Mind and the Endangered Origins of Intelligence*. Reading, MA: Addison-Wesley.

Gur, Rubin C., Lyn Harper, Susan M. Resnick, et al. 1995. "Sex Differences in Regional Cerebral Glucose Metabolism during a Resting State." *Science* 267: 528–31.

Hamer, D. H., S. Hu, V. L. Magnusen, et al. 1993. "A Linkage between DNA Markers on the X-Chromosome and Male Sexual Orientation." *Science* 261: 321–37.

Hamit, Francis. 1994. *Virtual Reality and the Exploration of Cyberspace*. Carmel, IN: SAMS.

Harth, Erich. 1993. *The Creative Loop: How the Brain Makes a Mind*. Reading, MA: Addison-Wesley.

Hilts, P. 1990. "U.S. Aides See Shaky Legal Basis for Ban on Fetal Tissue Research." *The New York Times* (January 30).

Hinton, Geoffrey E. 1992. "How Neural Networks Learn from Experience." *Scientific American* (September): 145–51.

Hoge, S. K., P. S. Appelbaum, T. Lawlor, et al. 1990. "A Prospective Multicenter Study of Patients' Refusal of Antipsychotic Medication." *Archives of General Psychiatry* 47: 949–56.

Holden, Constance. 1991. "Probing the Complex Genetics of Alcoholism." *Science* 251: 163–64.

Honderich, Ted. 1993. *How Free Are You? The Determinism Problem*. Oxford: Oxford University Press.

Hooper, Judith, and Dick Teresi. 1992. *The Three-Pound Universe*. New York: G.P. Putnam's Sons.

Horellou, Philippe, Frederic Revah, Oliver Sebate, et al. 1996. "Adenovirus: A New Tool to Transfer Genes into the Central Nervous System for Treatment of Neurodegenerative Disorders." In David S. Latchman, ed., *Genetic Manipulation of the Nervous System*. London: Academic Press.

Hotz, Robert Lee. 1997. "Estrogen May Play Key Role in Brain." *Los Angeles Times*, January 26.

Howard, Linda G. 1981. "Hazardous Substances in the Workplace: The Rights of Women." *University of Pennsylvania Law Review* 129: 798–845.

Hyde, Jane S., and Marcia C. Linn. 1988. "Gender Differences in Verbal Ability: A Meta-Analysis." *Psychological Bulletin* 104(1): 53–69.

Institute of Medicine. 1991. *Mapping the Brain and Its Functions: Integrating Enabling Technologies into Neuroscience Research*. Washington, D.C.: National Academy Press.

———. 1992. *Discovering the Brain*. Washington, D.C.: National Academy Press.

Interagency Council on the Homeless. 1992. *Outcast on Main Street: Report of the Federal Task Force on Homelessness and Severe Mental Illness*. Washington, D.C.: Government Printing Office, 92–1904.

Isaacson, Robert L., and Karl Jensen, eds. 1994. *The Vulnerable Brain and Environmental Risks*. London: Plenum.

Ivy, Gwen O. 1996. "The Aging Nervous System." In Andrious Baskys and Gary Remington, eds., *Brain Mechanisms and Psychotropic Drugs*. Boca Raton: CRC Press.

Jeffery, C. Ray. 1994. "The Brain, the Law, and the Medicalization of Crime." In Roger D. Masters and Michael T. McGuire, eds., *The Neurotransmitter Revolution: Serotonin, Social Behavior, and the Law*. Carbondale, IL: Southern Illinois University Press.

Joynt, Robert J. 1996. "Neurology." *JAMA* 275(23): 1826–27.

Kamil, Rifaat. 1996. "Antidepressants." In Andrious Baskys and Gary Remington, eds., *Brain Mechanisms and Psychotropic Drugs*. Boca Raton: CRC Press.

Kandel, Eric R., and Robert D. Hawkins. 1992. "The Biological Basis of Learning and Individuality." *Scientific American* (September): 79–86.

Kane, Robert. 1996. *The Significance of Free Will*. New York: Oxford University Press.

Keeton, Kathy, and Yvonne Baskin. 1985. *Women of Tomorrow*. New York: St. Martin's Press.

Kimura, Doreen. 1992. "Sex Differences and the Brain." *Scientific American* (September): 119–25.

Kosslyn, Stephen M., and Oliver Koenig. 1992. *Wet Mind: The New Cognitive Science*. New York: Free Press.

Kotulak, Ronald. 1996. *Inside the Brain: Revolutionary Discoveries of How the Brain Works*. Kansas City: Andrews and McMeel.

Krimsky, Sheldon, Ruth Hubbard, and Colin Gracey. 1988. "Fetal Research in the United States: A Historical and Ethical Perspective." *Gene Watch* 5(4/5): 1–3, 8–10.

Lamberts, Steven W. J., Annewieke W. van den Beld, and Aart-Jan van der Lely. 1997. "The Endocrinology of Aging." *Science* 278: 419–24.

Lappé, Marc. 1987. "The Limits of Genetic Inquiry." *Hastings Center Report* 17(4): 5–10.

Latchman, David S. 1996. "Genetic Manipulation of the Nervous System: An Overview." In David S. Latchman, ed., *Genetic Manipulation of the Nervous System*. London: Academic Press.

Leary, Warren E. 1997. "Fetal Tissue Put in Spine." *Sarasota Herald-Tribune*, July 13, A1, A16.

Lefley, H. P. 1992. "The Stigmatized Family." In P. J. Fink and A. Tasman, eds., *Stigma and Mental Illness*. Washington, D.C.: American Psychiatric Press.

Lemco, Jonathan. 1994. "Conclusion." In Jonathan Lemco, ed., *National Health Care: Lessons from the United States and Canada*. Ann Arbor: University of Michigan Press.

Lesch, Klaus-Peter, Dietmar Bengel, Armin Heils, et al. 1996. "Association of Anxiety-Related Traits with a Polymorphism in the Serotonin Transporter Gene Regulatory Region." *Science* 274: 1527–31.

Leshner, Alan I. 1997. "Addiction Is a Brain Disease, and It Matters." *Science* 278: 45–47.

LeVay, Simon. 1991. "A Difference in Hypothalamic Structure between Heterosexual and Homosexual Men." *Science* 253: 1034–37.

———. 1994. *The Sexual Brain*. Cambridge: MIT Press.

Levin, S., D. Yurgelun-Todd, and S. Craft. 1989. "Contributions of Clinical Neuropsychology to the Study of Schizophrenia." *Journal of Abnormal Psychology* 98: 341–56.

Lewis, D., et al. 1988. "Neuropsychiatric, Psychoeducational, and Family Characteristics of 14 Juveniles Condemned to Death in the United States." *American Journal of Psychiatry* 145(5): 584–89.

Lindvall, Olle, Patrik K. Brundin, Hakan Widner, et al. 1990. "Grafts of Fetal Dopamine Neurons Survive and Improve Motor Function in Parkinson's Disease." *Science* 247: 574–77.

Link, B. G., F. T. Cullen, J. Minotznik, et al. 1992. "The Consequences of Stigma for Persons with Mental Illness: Evidence from the Social Sciences." In P. J. Fink and A. Tasman, eds., *Stigma and Mental Illness*, Washington, D.C.: American Psychiatric Press.

Lois, Carlos, and Arturo Alvarez-Buyllo. 1996. "Neuronal Precursors in the Brain of Adult Mammals: Biology and Applications." In David S. Latchman, ed., *Genetic Manipulation in the Nervous System*. London: Academic Press.

Lomas, J., and A. P. Contandriopoulos. 1994. "Regulating Limits to Medicine: Towards Harmony in Public- and Self-Regulation." In Theodore R. Marmor, ed., *Understanding Health Care Reforms*. New Haven: Yale University Press.

Lowenstein, Pedro R., Gavin W. G. Wilkinson, M. G. Castro, et al. 1996. "Non-Neurotrophic Adenovirus: A Vector for Gene Transfer to the Brain and Possible Gene Therapy of Neurological Disorders." In David S. Latchman, ed., *Genetic Manipulation of the Nervous System*. London: Academic Press.

Lykken, David. 1997. "Incompetent Parenting: Its Causes and Cures," *Child Psychology and Human Development* 27, 3: 129–37.

Mahowald, Mary B. 1993. "Brain Development, Identity, and Neurografts." In Robert H. Blank and Andrea L. Bonnicksen, eds., *Emerging Issues in Biomedical Policy*. Vol. II. New York: Columbia University Press.

Marshall, E. 1995. "NIH 'Gay Gene' Study Questioned." *Science* 268: 1841.

Masters, Roger D. 1994. "Why Study Serotonin, Social Behavior, and the Law?" In Roger D. Masters and Michael T. McGuire, eds., *The Neurotransmitter Revolution: Serotonin, Social Behavior, and the Law*. Carbondale, IL: Southern Illinois University Press.

———. 1996. "Neuroscience, Genetics, and Society: Is the Biology of Human Social Behavior Too Controversial to Study?" *Politics and the Life Sciences* 15(1):103–4.

Masters, Roger D., and Michael T. McGuire, eds. 1994. *The Neurotransmitter Revolution: Serotonin, Social Behavior and the Law*. Carbondale, IL: Southern Illinois University Press.

Mechanic, David. 1994. *Inescapable Decisions: The Imperatives of Health Care*. New Brunswick, NJ: Transaction.

Miller, Jonathan. 1992. "Trouble in Mind." *Scientific American* (September): 180.

Mirsky, A. F. and M. H. Orzack. 1977. "Final Report on Psychosurgery Pilot Study." In *Appendix: Psychosurgery*. Washington, D.C.: Government Printing Office.

Monahan, J. 1992. "Mental Disorder and Violent Behavior: Perceptions and Evidence." *American Psychologist* 47: 511–21.

Morgall, Janine Marie. 1993. *Technology Assessment: A Feminist Perspective*. Philadelphia: Temple University Press.

Morrison, John H., and Patrick R. Hof. 1997. "Life and Death of Neurons in the Aging Brain." *Science* 278: 412–18.

Moss, D. C. 1988. "Ritalin Under Fire: 16 Lawsuits Claim Drug Was Wrongly Prescribed." *American Bar Association Journal* 74: 19.

Moussa, Mario, and Thomas A. Shannon. 1992. "The Search for the New Pineal Gland: Brain Life and Personhood." *Hastings Center Report* 22(2): 30–37.

Nadeau, Robert L. 1996. *S/HE Brain: Science, Sexual Politics and the Myths of Feminism*. Westport, CT: Praeger Press.

National Advisory Neurological and Communicative Disorders and Stroke Council. 1989. *Decade of the Brain: Answers through Scientific Research*. Washington, D.C.: GPO.

National Commission for the Protection of Human Subjects of Biomedical and Behavioral Research. 1976. *Psychosurgery*. Washington, D.C.: Government Printing Office.

National Institute of Neurological Disorders and Stroke. 1995. "Tissue Plasminogen Activator for Acute Ischemic Stroke." *New England Journal of Medicine* 333: 1632–33.

National Institute on Aging. 1993. *Progress Report on Alzheimer's Disease 1993*. Bethesda, MD: National Institutes of Health.

National Institutes of Health. 1988b. *Human Fetal Transplantation Research*. Bethesda, MD: NIH.

———. 1988a. *Report of Research Panel on Human Fetal Issue Transplantation*. Bethesda, MD: NIH.

National Organization on Disability. 1991. "Public Attitudes toward People with Disabilities." Survey conducted by Louis Harris and Associates, Inc.

National Research Council. 1992. *Environmental Neurotoxicology*. Washington, D.C.: National Academy Press.

Neergaard, Lauran. 1997. "FDA OKs 'Deep Brain Stimulator.'" *Sarasota Herald-Tribune*, August 5, B4.

Nelkin, Dorothy, and M. Susan Lindee. 1995. *The DNA Mystique: The Gene as a Cultural Icon*. New York: W. H. Freeman.

Nestler, Eric J., and George K. Aghajanian. 1997. "Molecular and Cellular Basis of Addiction." *Science* 278: 58–63.

Neuwelt, Edward A., et al. 1995. "Gene Replacement Therapy in the Central Nervous System: Viral Vector-Mediated Therapy of Global Neurodegenerative Disease." *Behavioral and Brain Sciences* 18: 1–9.

Nicholson, C. D. 1989. "Nootropics and Metabolically Active Compounds in Alzheimer's Disease." *Biochemical Society Transactions* 17(10): 83–85.

Office of Technology Assessment. 1976. *Development of Medical Technology: Opportunities for Assessment*. Washington, D.C.: Government Printing Office.

———. 1988. *Biology, Medicine and the Bill of Rights*. Washington, D.C.: GPO.

———. 1990. *Neural Grafting: Repairing the Brain and Spinal Cord*. Washington, D.C.: GPO.

———. 1990. *Neurotoxicity: Identifying and Controlling Poisons of the Nervous System*. Washington, D.C.: GPO.

———. 1992. *The Biology of Mental Disorders*. Washington, D.C.: GPO.

Ormel, J., W. Van der Brink, M. W. Koeter, et al. 1990. "Recognition, Management and Outcome of Psychological Disorders in Primary Care: A Nationalistic Follow-up Study." *Psychological Medicine* 20: 909–23.

Parens, Eric. 1996. "Taking Behavioral Genetics Seriously." *Hastings Center Report* 26(4): 13–18.

Petersen, Glen N. 1994. "Regulation of Electroconvulsive Therapy: The California Experience." In Harold I. Schwartz, ed., *Psychiatric Practice under Fire*. Washington, D.C.: American Psychiatric Press.

Polymeropoulos, Michael H., Christian Lavedan, Elisabeth Leroy, et al. 1997. "Mutation in the a-Synclein Gene Identified in Families with Parkinson's Disease." *Science* 276: 2045–49.

Pope, Harrison G., and Deborah Yurgelun-Todd. 1996. "The Residual Cognitive Effects of Heavy Marijuana Use in College Students." *Journal of the American Medical Association* 275(7): 521–27.

President's Commission for the Study of Ethical Problems in Medicine and Biomedical and Behavioral Research. 1981. *Defining Death*. Washington, D.C.: Government Printing Office.

Prince, Alan, and Paul Smolensky. 1997. "Optimality: From Neural Networks to Universal Grammar." *Science* 275: 1604–10.

Randall, Teri. 1993. "Demographers Ponder the Aging of the Aged and Await Unprecedented Looming Elder Boom." *Journal of the American Medical Association* 269(18): 2330–31.

Restak, Richard M. 1994a. *The Modular Brain*. New York: Simon and Schuster.

————. 1994b. *Receptors*. New York: Bantam Books.

Rickels, K., and E. Schweizer. 1990. "Clinical Overview of Serotonin Reuptake Inhibitors." *Journal of Clinical Psychiatry* 51: 9–12.

Roach, Gary W., Marc Kanchuger, Christina Mora Mangano, et al. 1996. "Adverse Cerebral Outcomes after Coronary Bypass Surgery." *New England Journal of Medicine* 335(25): 1857–63.

Robinson, Bambi E. S. 1993. "The Moral Permissibility of *In Utero* Experimentation." *Women and Politics* 13(3/4): 19–30.

Rodriguez de Fonseca, Fernando, M. Rocio, A. Carrera, et al. 1997. "Activation of Corticotropin-Releasing Factor in the Limbic System during Cannabinoid Withdrawal." *Science* 276: 2050–54.

Rose, S. 1993. "No Way to Treat the Mind." *New Scientist* 1869: 23–26.

Rosenbaum, Jerrold F. 1994: "Clinical Trial by Media: The Prozac Story." In Harold I. Schwartz, ed., *Psychiatric Practice under Fire*. Washington, D.C.: American Psychiatric Press.

Roses, Allen D. 1995. Testimony at Hearing before Special Committee on Aging, U.S. Senate, 104th Congress, 2nd Session, June 27. *Breakthroughs in Brain Research*. Washington, D.C.: GPO.

Safer, Daniel J., and John M. Krager. 1992. "Effect of a Media Blitz and a Threatened Lawsuit on Stimulant Treatment." *Journal of American Medical Association* 268 (8): 1004:1007.

Salmon, Paul G. 1990. "A Psychological Perspective on Musical Performance Anxiety: A Review of the Literature." *Medical Problems of Performing Artists* (March): 2–11.

Schuklenk, Udo, Edward Stein, Jacinta Kerin, and William Byne. 1997. "The Ethics of Genetic Research on Sexual Orientation." *Hastings Center Report* 27(4): 6–13.

Schwartz, Harold I., et al. 1988. "Autonomy and the Right to Refuse Treatment: Patients' Attitudes after Involuntary Medication." *Hospital and Community Psychiatry* 39: 1049–54.

"Science Does it With Feeling." 1996. *Economist* (20 July): 75–77.

Scott, Alwyn. 1995. *Stairway to the Mind: The Controversial New Science of Consciousness*. New York: Copernicus.

Seidenberg, Mark S. 1997. "Language Acquisition and Use: Learning and Applying Probabilistic Constraints." *Science* 275: 1599–1603.

Seiger, A., A. Nordberg, H. von Holst, et al. 1993. "Intracranial Infusion of Purified Nerve Growth Factor to an Alzheimer's Patient." *Behavior and Brain Research* 57(2): 255–61.

Selkoe, Dennis J. 1992. "Aging Brain, Aging Mind." *Scientific American* (September): 135–42.

————. 1995. Testimony at Hearing before Special Committee on Aging, U.S. Senate, 104th Congress, 2nd Session, June 27. *Breakthroughs in Brain Research*. Washington, D.C.: GPO.

Sena-Esteves, Miquel, Manish Aghi, Peter A. Pechan, et al. 1996. "Gene Delivery to the Nervous System Using Retroviral Vectors." In David S. Latchman, ed., *Genetic Manipulation of the Nervous System*. London: Academic Press.

Senut, Marie-Claude, Steven T. Suhr, and Fred H. Gage. 1996. "Transplantation of Genetically Modified Non-Neuronal Cells in the Central Nervous System." In David S. Latchman, ed., *Genetic Manipulation of the Nervous System*. London: Academic Press.

Shapiro, Michael H. 1994. "Law, Culpability, and the Neural Sciences." In Roger D. Masters and Michael T. McGuire, eds., *The Neurotransmitter Revolution: Serotonin, Social Behavior, and the Law*. Carbondale, IL: Southern Illinois University Press.

Shatz, Carla J. 1992. "The Developing Brain." *Scientific American* (September): 61–67.

Shaywitz, Bennett A., Sally E. Shaywitz, Kenneth R. Pugh, et al. 1995. "Sex Differences in the Functional Organization of the Brain for Language." *Nature* 373: 607–9.

Sheldon, Jan. 1987. "Legal and Ethical Issues in the Behavioral Treatment of Juvenile and Adult Offenders." In Edward Morris and Curtis Braukmann, eds., *Behavioral Approaches to Crime and Delinquency*. New York: Plenum Press.

Shine, H. David, and Savio L. C. Woo. 1996. "Adenovirus-Mediated Gene Therapy of Tumors in the Central Nervous System." In David S. Latchman, ed., *Genetic Manipulation of the Nervous System*. London: Academic Press.

Slomka, Jacquelyn. 1992. "Playing with Propranolol." *Hastings Center Report* 22(4): 13–17.

Smith, J. M. 1977. "Congenital Minimata Disease: Methyl Mercury Poisoning and Birth Defects in Japan." In E. Bingham, ed., *Proceedings: Conference on Women in the Workplace*. Washington, D.C.: Society for Occupational and Environmental Health.

Spencer, Dennis D., Richard J. Robbins, Frederick Naftolin, et al. 1992. "Unilateral Transplantation of Human Fetal Mesencephalic Tissue into the Candate Nucleus of Patients with Parkinson's Disease." *New England Journal of Medicine* 327(22): 1541–48.

Staver, Sari. 1989. "ECT Stigma Remains, Despite Effectiveness, Panel Says," *American Medical News* (May 26): 12.

Stein, Donald G., and Marylou M. Glasier. 1995. "Some Practical and Theoretical Issues Concerning Fetal Brain Tissue Grafts as Therapy for Brain Dysfunctions." *Behavioral and Brain Sciences* 18: 36–45.

Stone, Richard. 1992. "Molecular 'Surgery' for Brain Tumors." *Science* 256: 1513.

Sullivan, Louis W., Secretary of the Department of Health and Human Services, Press Release, November 2, 1989.

Szasz, Thomas. 1974. *The Myth of Mental Illness*. New York: Harper and Row.

Tanda, Gianluigi, Francesco E. Pontieri, and Gaetano DiChiara. 1997. "Cannabinoid and Heroin Activation of Mesolimbic Dopamine Transmission by a Common u1 Opioid Receptor Mechanism." *Science* 276: 2048–50.

Teuber, H. L., S. H. Corkin, and T. E. Twitchell. 1977. "Study of Cingulotomy in Man: A Summary." In W. H. Sweet, et al., eds. *Neurosurgical Treatment in Psychiatry, Pain and Epilepsy*. Baltimore: University Park Press.

Traumatic Brain Injury Act. 1993. U.S. Senate, 103rd Congress, 2nd Session, Report 103–243.

Truog, Robert D. 1997. "Is It Time to Abandon Brain Death?" *Hastings Center Report* 27(1): 29–37.

U.S. Congress, Senate Committee on Labor and Human Resources. 1994. *Traumatic Brain Injury Act of 1993 Report*. 103rd Congress, 2nd Session. Washington, D.C.: GPO.

U.S. Congress, Senate Special Committee on Aging. 1995. *Breakthroughs in Brain Research: A National Strategy to Save Billions in Health Care Costs*. 104th Congress, 1st Session. Washington, D.C.: GPO.

U.S. Department of Health and Human Services, National Institute of Neurological Disorders and Stroke. 1989. *Interagency Head Injury Task Force Report*. Bethesda, MD: NIH.

U.S. Environmental Protection Agency. 1986. *Air Quality Criteria for Lead*. EPA Report Number 600/8-83/028a F-dF.4V.

Valenstein, Elliott S., ed. 1986. *Great and Desperate Cures: The Rise and Decline of Psychosurgery and Other Radical Treatments for Mental Illness*. New York: Basic Books.

Vawter, Dorothy E. 1993. "Fetal Tissue Transplantation Policy in the United States." *Politics and the Life Sciences* 12(1): 79–85.

Veatch, Robert M. 1993. "The Impending Collapse of the Whole-Brain Definition of Death." *Hastings Center Report* 23(4): 18–24.

Veggeberg, Scott. 1996. "Beyond Steroids." *New Scientist* (January 13): 28–31.

Wahl, O. F., and C. R. Harman. 1989. "Family Views of Stigma." *Schizophrenia Bulletin* 15: 131–39.

Waldo, Daniel R., et al. 1989. "Health Expenditures by Age Group; 1977 and 1987." *Health Care Financing Review* 10: 116–18.

Weigele, Thomas C. 1979. *Biopolitics: A Search for a More Human Political Science*. Boulder: Westview Press.

Weiss, Bernard. 1985. Testimony before the Committee on Science and Technology. U.S. Congress, House of Representatives, October 8.

Weiss, Bernard, and J. L. O'Donoghue, eds. 1994. *Neurobehavioral Toxicology: Analysis and Interpretation*. New York: Raven Press.

Whalley, L. J. 1993. "Ethical Issues in the Application of Virtual Reality to the Treatment of Mental Disorders." In R. A. Earnshaw, M. A. Gigante, and H. Jones, eds., *Virtual Reality Systems*. London: Academic Press.

White, Elliott. 1992. *The End of the Empty Organism: Neurobiology and the Sciences of Human Action*. Westport, CT: Praeger.

Whitehouse, Peter J., Eric Juengst, Maxwell Mehlman, and Thomas H. Murray. 1997. "Enhancing Cognition in the Intellectually Intact." *Hastings Center Report* 27(3): 14–22.

Whitfield, Charles L. 1995. *Memory and Abuse: Remembering and Healing the Effects of Trauma*. New York: Health Communications, Inc.

Wickelgren, Ingrid. 1997a. "Getting a Grasp on Working Memory." *Science* 275: 1580–82.

———. 1997b. "Marijuana: Harder than Thought?" *Science* 276: 1967–68.

Wikler, Daniel. 1994. "Predictive Testing in Genetics and Psychiatry: Ethical Issues in the Use of Advance Directives." In Norio Fujiki and Darryl Macer, eds.,

Intractable Neurological Disorders, Human Genome Research and Society. Tsukaba, Japan: Eubios Ethics Institute.

Williams, Christopher. 1996. *Environmentally-Mediated Intellectual Decline (EMID): A Selected Interdisciplinary Bibliography.* Cambridge: Cambridge University Global Security Programme.

Wills, Christopher. 1993. *The Runaway Brain: The Evolution of Human Uniqueness.* New York: Basic Books.

Wilson, Edward O. 1978. *On Human Nature.* Cambridge, MA: Harvard University Press.

Windisch, Manfred. 1996. "Cognitive-Enhancing (Nootropic) Drugs." In Andrious Baskys and Gary Remington, eds., *Brain Mechanisms and Psychotropic Drugs.* Boca Raton: CRC Press.

Winock, Bruce J. 1997. *The Right to Refuse Mental Health Treatment.* Washington, D.C.: American Psychological Association.

Wise, R. A. 1988. "The Neurobiology of Craving: Implications for the Understanding and Treatment of Addiction." *Journal of Abnormal Psychology.* 97:118–32.

Wojtowicz, J. Martin. 1996. "Membranes, Synapses, and Ion Channels." In Andrious Baskys and Gary Remington, eds., *Brain Mechanisms and Psychotropic Drugs.* Boca Raton: CRC Press.

Yesley, Michael S. 1994. "The Behaviour-Genetics Debate in the United States." In Norio Fujiki and Darryl Macer, eds., *Intractable Neurological Disorders, Human Genome Research and Society.* Tsukaba, Japan: Eubios Ethics Institute.

Zedi, Semir. 1992. "The Visual Image in Mind and Brain." *Scientific American* (September): 69–76.

Index